Lecture Notes in Electrical Engineering

Volume 1108

The book series *Lecture Notes in Electrical Engineering* (LNEE) publishes the latest developments in Electrical Engineering—quickly, informally and in high quality. While original research reported in proceedings and monographs has traditionally formed the core of LNEE, we also encourage authors to submit books devoted to supporting student education and professional training in the various fields and applications areas of electrical engineering. The series cover classical and emerging topics concerning:

- Communication Engineering, Information Theory and Networks
- Electronics Engineering and Microelectronics
- Signal, Image and Speech Processing
- Wireless and Mobile Communication
- Circuits and Systems
- Energy Systems, Power Electronics and Electrical Machines
- Electro-optical Engineering
- Instrumentation Engineering
- Avionics Engineering
- Control Systems
- Internet-of-Things and Cybersecurity
- Biomedical Devices, MEMS and NEMS

For general information about this book series, comments or suggestions, please contact leontina.dicecco@springer.com.

To submit a proposal or request further information, please contact the Publishing Editor in your country:

China

Jasmine Dou, Editor (jasmine.dou@springer.com)

India, Japan, Rest of Asia

Swati Meherishi, Editorial Director (Swati.Meherishi@springer.com)

Southeast Asia, Australia, New Zealand

Ramesh Nath Premnath, Editor (ramesh.premnath@springernature.com)

USA, Canada

Michael Luby, Senior Editor (michael.luby@springer.com)

All other Countries

Leontina Di Cecco, Senior Editor (leontina.dicecco@springer.com)

**** This series is indexed by EI Compendex and Scopus databases. ****

Martin Krchňák · Marek Češkovič · Pavol Kurdel ·
Anton Panda

Advancements in Antenna Measurement

A Novel Approach to High-Frequency Attenuation

Springer

Martin Krchňák
Department of Avionics
Technical University of Košice
Košice, Slovakia

Marek Češkovič
Department of Avionics
Technical University of Košice
Košice, Slovakia

Pavol Kurdel
Department of Avionics
Technical University of Košice
Košice, Slovakia

Anton Panda
Department of Automobile
and Manufacturing Technologies
Technical University of Košice
Prešov, Slovakia

ISSN 1876-1100 ISSN 1876-1119 (electronic)
Lecture Notes in Electrical Engineering
ISBN 978-3-031-48834-4 ISBN 978-3-031-48835-1 (eBook)
https://doi.org/10.1007/978-3-031-48835-1

This Springer imprint is published by the registered company Springer Nature Switzerland AG
The registered company address is: Gewerbestrasse 11, 6330 Cham, Switzerland

Paper in this product is recyclable.

Preface

The presented book aims to find a new form of the high-frequency anechoic chamber, which could, in a specific limited form, replace the classical non-reflection anechoic chamber. Such a chamber could allow one to measure the directional characteristics of the antennas. The motivation is a new idea, representing the use of the classical principle of changing the polarization of radio waves in a form without a reflection chamber. The classical non-reflective chamber uses the principle of absorbing electromagnetic waves with a damping material that covers the inner walls of the closed and limited space of the damping chamber. The absorption material will prevent the occurrence of parasitic reflections from the solid walls of the chamber and thus prevent the measurement of the directional properties of the antenna from being disturbed. Another well-known principle is the measurement of antennas in free space. If it is free wide and far from the measuring workplace, e.g., external space, without reflective surfaces or metal materials, there are also no parasitic reflections of HF energy. While the chamber is a relatively expensive device in the first case, the measuring workplace is space intensive in the second case. For an ordinary worker in the antenna technology field, the antenna measurement could be carried out, e.g., in a classical laboratory. However, such a laboratory is full of various reflective surfaces and metal devices that do not allow the measurement of radiation characteristics due to parasitic reflections. Moreover, it is in such an environment that it is possible to use the principle of changing the polarization of radio waves as a non-reflective environment for measuring the directional properties of antennas. For this purpose, the presented work is devoted to analysing the experimental results of the measurement and searching for the optimal shape and composition of the depolarization panel, which would fulfil such a role when measuring antennas. The approach to the processing of the publication was also chosen for this direction. On the one hand, the methodology of frequency analysis and optimisation of the depolarization panel is presented, presented in the fifth chapter. On the other hand, it is the methodology of finding the optimal layout—creating a measuring workplace using a depolarization panel, presented in the sixth chapter. The general description of the investigated problem can be characterised as the search for a suitable design of the depolarization panel and its placement during antenna measurement in space so that between the transmitting and receiving antennas, it suppresses—attenuates the reflected parasitic signal of the same polarization. To highlight the direct link between the transmitting and receiving antenna through the depolarization panel, i.e., by changing the polarization. Thus, the suppression of the cross-polarization coupling between the transmitting and receiving antenna is the primary attenuation phenomenon that will allow the measurement of the directional properties of antennas in any laboratory environment. The last seventh chapter of the publication presents the encouraging results of applying this measurement method in antenna technology.

Košice, Slovakia Martin Krchňák
Košice, Slovakia Marek Češkovič
Košice, Slovakia Pavol Kurdel
Prešov, Slovakia Anton Panda

Acknowledgments Author expresses gratitude to reviewers for valuable book, substantive and formal observations raise the overall level of quality publications.

Publisher/editor: Springer International Publishing, Switzerland

Reviewers:
Prof. Luk Kwai-man
Prof. Colin Fidge
Prof. Ing. Andrej Novák

Contents

About the Authors

Martin Krchňák, Ing., Ph.D. has experience and research interests in radio reflection and attenuation of construction materials, aircraft radio devices and aircraft antenna prototypes, new antenna measurement principles and measures.

Marek Češkovič, Ing., Ph.D. is a member of the research team of the laboratories of Aircraft antenna measurements and Laboratory of Intelligent Control Systems of Aircraft Engines (LIRSLM). Since 2014 he works as pedagogue and scientist at the Faculty of Aeronautics of Technical University of Košice.

Assoc. Prof. Pavol Kurdel, Ing., Ph.D. is manager of quality system management on faculty of Aerospace, Technical University in Košice, Slovakia. His scientific interests include electronic aircraft systems, navigation aviation systems, airport processes, and remote sensing.

Prof. Anton Panda, Ing., Ph.D. is a pedagogue and scientist at the Faculty of manufacturing technologies TU Košice with the seat in Prešov, as well as auditor of quality management systems. He has research interests in automobile production, manufacturing technologies, experimental methods in the manufacturing technologies, machining, development, manufacturing among others. He is a member of the Polish Academy of Sciences and of ASME, USA.

Symbols and Abbreviations

3D	Three—Dimensional
AC	Automatic computer
AM	Amplitude modulation
Annex	Standard of military and government
Bit	Binary digit, the name of the basic unit of information
CAD-FEKO	(Feldberechnung bei Korpern mat beliebiger Oberflache), field computations involving bodies of arbitrary shape, hybrid solver of MoM and UTD/PO
CPU	Central Processing Unit
dB	Decibel, expresses the ratio of two values of a power or root-power quantity on a logarithmic scale
dBm	Decibels compared with 1 milliwatt.
DC	Direct Current
DO	Digital output
DAQ	Data acquisition, high sampling rate data acquisition device
DUT port	A device under test, model with S parameters given by the square of the number of ports
E	Symbol for electric field
f	Symbol for frequency in Hz
FM	Frequency Modulation
HF	High frequency for the range of radio frequency between 3 and 30 MHz
H	Symbol for magnetic field intensity
IEEE	Professional association for electronics engineering, electrical engineering, and other related disciplines
I	Symbol for current
I/O	Input-Output (Device)
LRL	Line reflect line
LabVIEW	Laboratory Virtual Instrumentation Engineering Workbench
λ	Wavelength is the distance between the repeating periods of the wave

m	Milli (prefix) -10^{-3}
MHz	Megahertz
Matlab	Programming and numeric computing platform used by millions of engineers and scientists to analyze data, develop algorithms, and create models
μ	Micro (prefix) 10^{-6}
OS	Operation system
PE	Polyethylene
PWR	Power
PC	Personal Computer
Port	Pair of terminals connecting an electrical network or circuit to an external circuit, as a point of entry or exit for electrical energy
Rx	Receiver
RF	Radio Frequency
RJ 45	Standard electrical connector for connecting cables
R&D	Research and Development
R/T	Real Time
rad	Radian (plane angle)
S—parameters	Derived from the "scattering" matrices described in the 1965 IEEE article "Power Waves and the Scattering Matrix"
s	Second
SB	Sideband
S/N	Serial number
SA	Spectrum analyzer
SOLT	Short open load thru, most used for calibration of coaxial systems
SSLT	Short-short load thru the most common algorithm for calibration of the system for measurements on waveguides
SSST	Short, short, short thru, an algorithm suitable for measurements of coaxial conduction and waveguides
Tx	Transmitter
U	A physical quantity that expresses the difference in the electric potential of two points and represents the energy required to move the electric charge between these two points in a certain electric field
USB	Upper Sideband
VSWR	Voltage Standing Wave Ratio, function of the reflection coefficient, which describes the power reflected from the antenna
WiFi	Wireless fidelity, wireless network protocols

Chapter 1
Introduction

The high-frequency anechoic chamber is an essential piece of equipment in the aerospace development laboratory environment. Its design creates an almost ideal measurement environment for RF technology. It not only measures the radiation characteristics of aircraft antennas but also provides space for measuring electromagnetic compatibility and the system's resistance to interference from the external electromagnetic field. However, the price of such an anechoic chamber ranges from several hundred thousand to one million euros. This fact is the main motivation for writing the monograph. It is the search for a different method, less costly than standard laboratory conditions, to obtain such antenna-specific radiation that would be comparable to the results of measurements in the anechoic chamber. This method of measuring characteristics has been the subject of research at the Faculty of Aeronautics of the Technical University of Košice for several years. Long-term scientific research activities have created a scenario focused on a viable alternative to the problem. In order to verify the final hypothesis, scientific research activity was started under the name "Application of the depolarization method in antenna technology". The aim of the research was to analyse and propose a new method of measuring the radiation characteristics of antennas that had not yet been presented in the scientific literature.

Currently, a significant amount of attention is given to the issue of measuring the radiation properties of antennas. By measuring the characteristics of the antenna system, it is possible to evaluate the qualitative properties of newly developed systems. A standardized type of antenna measuring technique requires the use of an anechoic chamber. Anechoic chambers vary in dimensions, frequency, and properties. The largest anechoic chambers in the world can take measurements directly on large commercial aircraft, such as Airbus A320 or Boeing 747. The smaller anechoic chambers are referred to as compact. These types of chambers find wide applicability in the automotive industry and other domains. The presented work entitled "Application of the depolarization method in antenna technology" addresses the design of a special quasi-"anechoic chamber". During the investigation, several important subset problems had to be addressed. The resulting measurement methodology

M. Krchňák et al., *Advancements in Antenna Measurement*, Lecture Notes in Electrical Engineering 1108, https://doi.org/10.1007/978-3-031-48835-1_1

was experimentally validated. As the presented method is original and has not been published before, appropriate literary sources are scarce in this instance. Significant publications on the interaction of electromagnetic waves relevant to this application include: [1–10] from the hypothesis under consideration in the above literature, it was feasible to develop a theoretical model and the notion of a depolarization panel with the intended characteristics. We conducted simulations based on 3D modelling to examine some selected features. Using the knowledge gained, a depolarization panel developed, and it underwent intensive testing. Then, we used this panel to conduct experimental measurements that focused on the radiating characteristics of specific antenna types.

This submission comprises seven chapters arranged chronologically in accordance with the methodical solution of the work. The final chapters of this monograph provide the most significant contribution to science. Chapter Five, titled "Optimization of the depolarization panel", thoroughly explains the principles of the depolarization panel and analyses the results of its optimization. Chapter six is titled "Optimization of the distribution of measurement elements" and discusses the thorough search procedure that resulted in the optimal layout of the workplace elements required for the analysis of antenna characteristics. Both of these chapters formed the basis for filing patent applications. The seventh and final chapter provides a comparative summary of directional characteristic measurements on the same antenna using two different methods. The first is the traditional method used worldwide, and in contrast, the monograph presents the depolarization method.

The presented method for directional characteristic measurements of antennas is groundbreaking because it enables obtaining similar results by measuring antennas' characteristics in a standard laboratory as one would achieve while measuring the same characteristics in an anechoic chamber using the conventional method.

References

1. Černohorsky, D., Tichý, J.: Vyzařování a šíření rádiových vln a antény. Brno (1977)
2. Davidov, P.: Rádionavigačné systémy lietadiel. Moskva (1980)
3. Dobeš, J., Žalud, V.: Moderní rádiotechnika. Praha (2006)
4. Garg, R.: Analytical and computational methods in electromagnetic (2008). ISBN 9781596933859
5. Hertz, H.: Electric waves, book on demand Ltd (2013). ISBN-13: 978-5518691582.
6. Joel, R.H.: Basic antennas – practical antennas and design. Amateur radio, USA (1994)
7. Clayton, P.: Introduction to electromagnetic compatibility, 2nd edn. (2006). ISBN 9780471755005
8. Prokop, J. – Vokurka, J.: Šíření elektromagnetických vln a antény. Praha 1982
9. Vavra, Š., Turán, J.: Antény a šírenie elektromagnetických vĺn. Bratislava (1989). ISBN 80-05-00131-2
10. Žalud, V.: Moderní rádioelektronika. Praha (2000)

Chapter 2
Anechoic Chambers for Radio Waves

This chapter deals with the current state of the art in the field of radio-frequency anechoic chambers used in domestic and foreign laboratories. The chapter describes the basic types of RF attenuation chambers and their construction. It also describes the different types of commonly used RF absorbers and the brief principle of such absorbers. The chapter presents test standards that require measurements to be made in anechoic chambers. At the end of the chapter, the timeliness of the topic is illustrated, as considerable funds are invested in the development of anechoic chambers, while the chambers are required for a wide range of measurements on different equipment. Therefore, it would be advantageous to build a new chamber that uses a different attenuation principle than conventional chambers, that is more affordable, or that is mobile to allow measurements in the field [1–4].

It is necessary to know the effects of the environment on each HF application. An application where it is necessary to know the influence of the environment is known is the measurement of HF devices [5]. These include, for example:

- Measurement of antenna characteristics,
- The influence of the object on the functionality of the antenna system,
- Measurement of interfering electromagnetic radiation from the device,
- Measurement of the near and distant fields of the antenna,
- Measurement of reflection and attenuation of various objects and materials,
- Measurement of radiation performance,
- Measurement of the immunity of the device to the HF signal.

The task of anechoic chambers is to protect the internal space from external interference and ensure that the wave that falls on the wall of the chamber is absorbed or reflected away from the measuring apparatus. The measurement space can be divided in several ways [6].

- full anechoic chamber–full anechoic chamber,
- measuring the workplace in open space,
- semi-anechoic chamber–semi-anechoic chamber.

M. Krchňák et al., *Advancements in Antenna Measurement*, Lecture Notes in Electrical Engineering 1108, https://doi.org/10.1007/978-3-031-48835-1_2

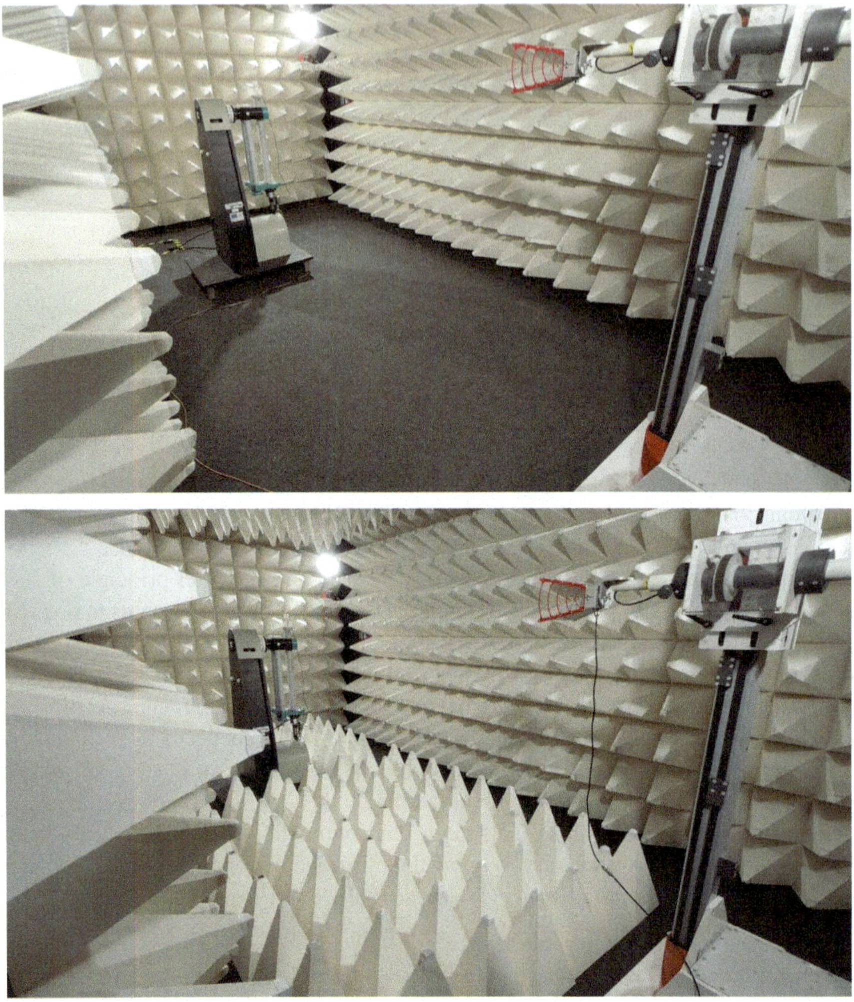

Fig. 2.1 Anechoic chamber, semi-anechoic chamber

The measurement space limitation described above applies to electromagnetic measurements and acoustic measurements [7]. Anechoic chambers can be used for low-frequency or high-frequency—radio waves. Next, it is necessary to consider the requirements for the signal's power, which will be used in the chamber and for the measured object's dimensions. The object's dimensions define the dimension of the entire chamber, which must be several times larger than the measured object. Finally, the so-called silence zone is defined in the chamber to ensure the best possible measurement properties, where the minimum energy of the reflected waves from the walls can be measured (Fig. 2.1).

Table 2.1 Testing standards [10]

Emission testing	Immunity testing
EN 61,000-6-3	
EN 61,000-6-4)	EN 61,000–6-1
CISPR 11/EN55011	EN 61,000-6-2
CISPR 12/EN55012	EN 55,020
CISPR 13/EN55013	EN 55,024
CISPR 14/EN55014	Mil-Std 461-F section HF
CISPR 15/EN55015	DO-160-G
CISPR 16-1-14	
CISPR 22/EN55022	
FCC part 15	
CISPR 25	

One of the uses of anechoic chambers is for radio measurements. Measurements involve the detection of very low levels of signals, and therefore a high level of isolation of the measurement area from the environment is required. Attenuation chambers create a space between the signal source and the DUT in which the parameters are defined. The properties of the artificially created space should be close to those of free space. At first glance, the design of a radio wave attenuation chamber may appear to be the same as that of an acoustic chamber. It also consists of walls on which absorbers are placed. They are not so different. Both acoustic and radio anechoic chambers are built to eliminate external influences on the measurement and to prevent retroreflection of the signal sent from the source to the target. However, the difference is significant, although not immediately obvious [8–11].

Measurements in the chambers are carried out according to the requirements of the standards. These define the permissible limits for RF emissions and the spurious signals in electromagnetic immunity tests (Table 2.1).

Radio chambers do not work with mechanical waves, like acoustic ones, but they work with radio waves. It means that the materials used for acoustic chambers will need to be more efficient for radio signals and vice versa. The most significant difference is the absorbers used. While only the porous material or the wall structure suffices for acoustic chambers, this issue is far more complex for radio waves. Therefore, unique materials can absorb radio waves over a wide range of frequencies and performances or reflect them so that they do not cause interference in the measurement zone [12–14].

2.1 The Principle of Attenuation in the Anechoic Chamber

At first glance, the principle of anechoic chambers may resemble a black box, but it is elementary. The theory of transmission lines is necessary to understand the principle of a reflective chamber. Let's suppose that a plane wave hits a wall. The wave impedance of the free space is 377 Ω [15]. If we use metal walls, the equivalent circuit will be represented by a transmission line terminated by a short. In such a circuit, no power is consumed by the load, so everything is reflected. However, if we wrap the wall with the material, it will represent a load on the wave. That is, the transmission line will not be terminated briefly but with a specific impedance corresponding to the impedance of the material with which we wrapped the walls (Fig. 2.2).

To demonstrate this type of material, the so-called Salisbury paper is used. This paper represents a plate of paper coated with a material with a surface resistance of 377 Ω/m^2. It is located precisely a quarter of the wavelength from the wall. The Salisbury paper is an elegant solution to the problem; however, its use is limited to only one frequency. Salisbury papers with different resistivity are used to achieve a wider frequency spectrum. Papers are placed sequentially, one after another, in multiples of a quarter of the wavelength from the wall. Such an arrangement reduces the reflection coefficient from 1 to less than 0.1, representing a difference in the reflected signal of more than 20 dB. The frequency range then changes from a single frequency to a range from 1 to 2.5 times the wavelength [15] (Fig. 2.3).

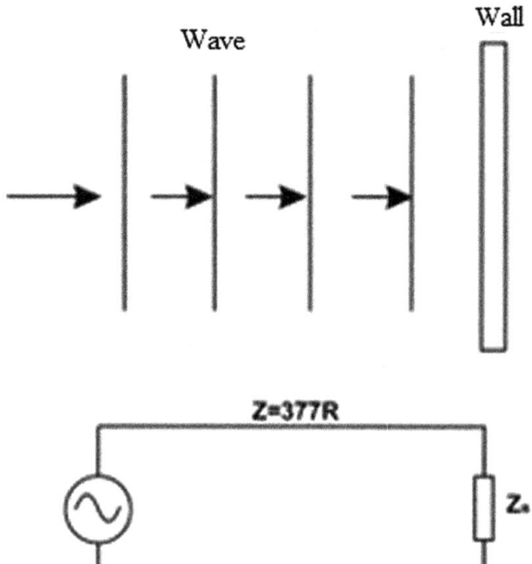

Fig. 2.2 Substitution scheme of a plane wave that strikes the wall [15]

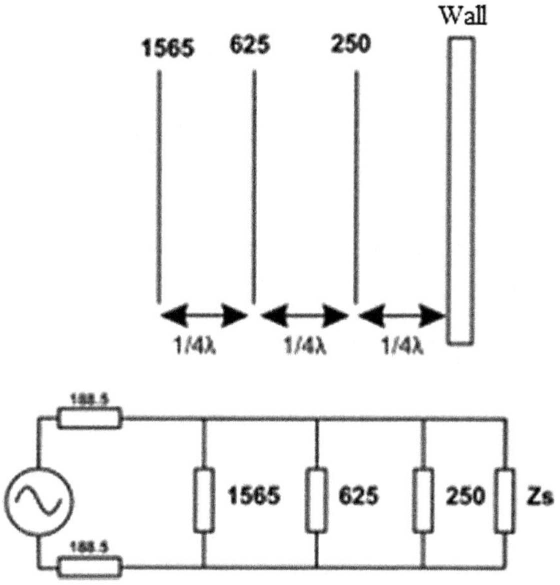

Fig. 2.3 Use of salisbury paper with different resistivity [15]

Another solution to such a problem is the so-called Jungmann sandwich. This arrangement gradually narrows the distance from the papers to the wall. The Jungmann sandwich reaches an attenuation of 20 dB at frequencies from 1 to 5 times the wavelength (Fig. 2.4).

Jungmann's sandwich principle is used by pyramid absorbers that imitate various impedances. When an electromagnetic wave falls on the pyramid, a series of small reflections are absorbed. Again, the principle of converting energy into heat under the influence of the material's strength applies. For pyramid absorbers to be effective, they must be at least half a wavelength of the minimum frequency long. From this

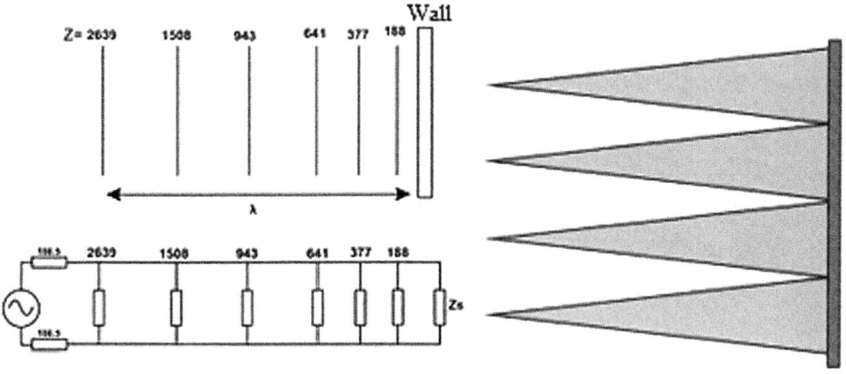

Fig. 2.4 Jungmann sandwich [15]

assumption, it is apparent that for frequencies below 100 MHz, the dimensions of the absorber would be enormous. For these frequencies, another material is used, namely, ferrite plates. The resistance of the ferrite plate is approximately equal to 377 Ω [15]. At 100 MHz, the attenuation of a ferrite plate 1 cm thick will be:

$$\mathrm{L} = \mathrm{e}^{-\mathrm{ad}} = \mathrm{e}^{-\frac{120\pi}{\lambda}(0.1)} = \mathrm{e}^{-1.26} = 11\ \mathrm{dB} \tag{2.1}$$

Thus, the wave that passes through the ferrite plate is subdued by 11 dB. In the case of a combination of ferrite plates and pyramidal absorbers, it is possible to assemble a chamber with an extensive range of frequencies.

2.2 The Construction of Anechoic Chambers

The anechoic chambers with a conductive floor, which we consider to be a perfectly conductive infinite ground plane, are called semi-anechoic chambers. The main purpose of these chambers is to measure the RF emissions generated by the equipment and to measure the immunity of the equipment to electromagnetic radiation. The frequency range of semi-anechoic chambers is 30–1000 MHz for emission measurements and 26–1000 MHz for immunity measurements. The frequency range is currently being extended to 6, 18 and 40 GHz. The frequency range is determined by the absorbers used, as it is not possible to produce a material that can operate in an arbitrarily wide frequency range at any signal power level (Fig. 2.5).

In most cases, two technologies of damping chambers and their combinations are used. The first is RF shielding and the second is the use of absorbers. The RF shielding technique is based on the principle of a shielded chamber according to the Faraday cage principle [16, 17], i.e. there is an isolated high-frequency environment inside the chamber that is not disturbed by external influences and, conversely, the signal source inside the chamber does not affect the external environment. Shielding

(a) (b)

Fig. 2.5 **a** Semi-anechoic chamber, **b** anechoic chamber [6] requirements range from 10 kHz to 18 GHz. These are based on the test methods described in IEEE 299 and EN 50,147-1. The walls of the chamber may consist of a single layer of steel sheet (single skin type) or several layers of metal (sandwich type), with the gaps between the sheets filled with damping material.

The measurement itself also affects all the components used in the chamber. It is therefore necessary to use components that have been approved by the manufacturer for use in such an environment. The component itself must have as little influence on the measurement process as possible, i.e., the component itself must not generate any interference or otherwise interfere with the measurement environment (protruding arms, metal and other conductive materials causing reflections, etc.). The measurement process is automated and controlled by a computer program, so that, for example, a robotic arm can be used to rotate the measured antenna in the case of antenna characteristic measurements, or, in the case of large chambers, an entire turntable can be used to rotate the entire aircraft or car in relation to a fixed measuring antenna. The electrical connection of these platforms must also not interfere with the measurement results, so optical fibres are used to transmit the signal [18–20].

2.3 HF Absorber of the Anechoic Chamber

The choice of absorbers is a key element in the design of the damping chamber. Absorbers must absorb as much of the signal as possible with as little reflection from the surface. Three types of absorbers are commonly used:

- o Microwave pyramid absorbers,
- o Ferrite plate,
- o Hybrid absorbers.

Microwave pyramid absorbers can be used from a frequency range of 200 MHz, in some cases as low as 80 MHz. The material is often referred to as blue foam and is a polyurethane foam with a carbon additive. To achieve attenuation, the size of the pyramid depends on the electrical parameters—the wavelength. For this reason, pyramid absorbers are most commonly used for high frequencies. Some chambers still use this material to measure electromagnetic compatibility, with pyramid absorbers over 2.4 m long achieving useful attenuation and reflection at less than 30 MHz. However, they are more commonly used in satellite and antenna-focusing chambers and have a frequency range of 500 MHz–40 GHz [12, 14].

Ferrite panels are a non-linear absorbing material. For them to work best, they must be mounted on a metal base, with the gap between the ferrite and the metal being critical. The plates have excellent attenuation properties at a relatively low height of 5–7 mm, but the attenuation drops significantly at frequencies of around 1.5 GHz and below. Manufacturers offer a compromise—plastic pyramid absorbers filled with ferrite powder. These absorbers are small and achieve attenuation of 15–20 dB up to 6 GHz and above. When used with ferrite plates, they must be tuned together. The use of ferrite plates increases the weight of the entire chamber, and it is necessary to use additional supports for the chamber, they are also expensive financially [12, 14].

Hybrid absorbers are created by combining both technologies. But it is not a combination of blue foam absorbers glued to the walls of ferrite. In this case, the ferrite function would be eliminated. The basic principle is that when a wave enters a material, it propagates through a medium with an impedance different from that of free space. The wave hits the ferrite plate, which has a different impedance to the absorber and bounces off the surface of the ferrite without passing through the ferrite, rendering the ferrite ineffective. The problem is solved by using a smaller amount of carbon or hollow pyramids [15] (Fig. 2.6).

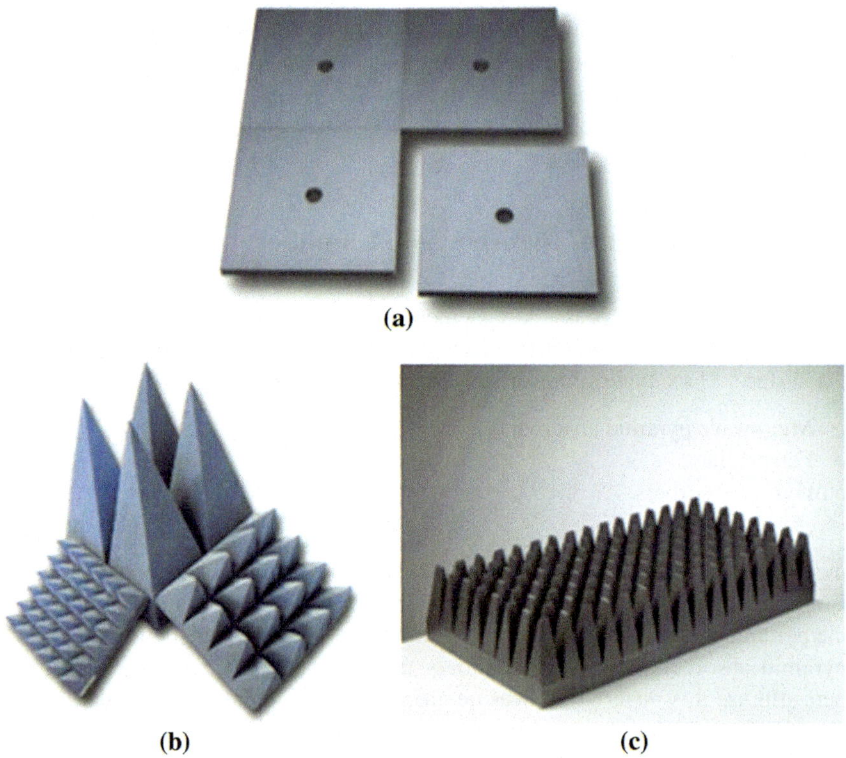

Fig. 2.6 **a** Ferrite plate, **b** foam absorber, **c** hybrid absorber [21, 22]

2.4 Important Laboratories Equipped with Anechoic Chambers

High-frequency anechoic chambers are costly pieces of equipment, take up a lot of space, and require expensive test accessories in the form of manipulators, test antennas, instrumentation and cabling. For this reason, such test facilities are designed for companies that are actively involved in RF R&D and may offer various services to their customers. One of the most common services offered is the testing and certification of equipment for electromagnetic compatibility. Such testing and certification services are provided by a number of companies. These can be local, for a particular region or country, or global, for the whole world. However, the standards to which they are tested are global, i.e. they apply to the whole world. Large local companies in the USA include Washington Laboratories, EMC Testing or CKC Laboratories, Inc. Much more common are global testing laboratories, which cover the whole world. The most important ones are UL, Intertek, Compliance Testing LLC, MET Labs and Technology International Inc. In Slovakia, such services are provided by the Electrotechnical Research and Design Institute and the Technical Testing Institute in Piestany.

The manufacturers of boxes usually offer boxes according to their use, which implies their dimensions. The most frequently offered chamber lengths are 3, 5 and 10 m. Such chambers are usually described as compact. Leading manufacturers of RF chambers include TDK RF Solutions, COMTEST Engineering, DJM Electronics, Frankonia Group and Panashield. Manufacturers are also able to design the chamber to meet the requirements of a specific purpose, e.g. testing of entire aircraft, cars, space or military equipment.

The largest full aircraft test chamber is the Benefield Anechoic Facility, which has tested full aircraft such as the B-52 Stratofortress, C-17 Globemaster, F-22 Raptor, C-130 Hercules, NC-130H, F-16 Fighting Falcon, B-1, Lancer X-43A, MH-47 Chinook, V-22 Osprey.

References

1. Joel, R.H.: Basic antennas – practical antennas and design. Amateur radio, USA (1994)
2. Labun, J.: Analýza chýb rádiovýškomerov malých výšok. Košice (1996)
3. Williams, Tim.: EMC for product designers, 4th edition (2007). ISBN 9780750681704
4. Anritsu corporation, Understanding VNA calibration Rev A, 06/2012, Anritsu Company (2012). ISBN 111410-00673A
5. Borghi, R., et al.: Plane waves catering by a set of perfectly conducting circular cylinders in the presence of a plane surface. J. Opt. Soc. Am. A **13**(12), 2441. https://doi.org/10.1364/josaa.13.002441
6. UHF Ferrite Absorber Tile (FAT-900).: Dostupné na internete: http://www.cfe.com.tw/2-absorber/crown-ferrite-absorber-ferrite-tile.jpg
7. S-parametre: popis, definícia a meranie. Dostupné na internete
8. Davidov, P.: Rádionavigačné systémy lietadiel. Moskva (1980)
9. Dobeš, J., Žalud, V.: Moderní rádiotechnika. Praha (2006)

10. Navedtra.: Antennas and wave propagation. Naval education and training Professional development and technology centre. ISBN 0504-LP-026-7580
11. Clayton, P.: Introduction to electromagnetic compatibility, 2nd edition (2006). ISBN 9780471755005
12. Anechoic Chambers/RF-Shielded Rooms. Dostupné na internete: https://www.reliantemc.com/download/FRANKONIA/Frankonia-Group-Anechoic-Chambers-2016.pdf.
13. Zubzanda, M.: Rádiové vybavenie lietadiel. Košice (1986)
14. Basic Rules for Anechoic Chamber Design, Part One: RF Absorber Approximations. Dostupné na internete. https://www.Microwave journal.com/articles/print/25704-basic-rules-for-anechoic-chamber-design-part-one-rf-absorber-approximations
15. Of the institution of electrical engineers—Part IIIA: radiolocation **93**(4), 620–626. https://doi.org/10.1049/ji-3a-1.1946.0150DASH
16. Besser, L., Gilmore, R: Practical RF circuit design for modern wireless systems. Vol. I: Passive Circuits Syst. (2003). ISBN 1580535216
17. Davidson, D.: Computational electromagnetic for RF and microwave engine erring (2005). ISBN 9780521070126.
18. Modelling and analysing polarization. Dostupné na internete: https://www.mathworks.com/help/phased/examples/modeling-and-analyzing-polarization.html
19. ÚVOD: Význam antén a ich základné parametre. Na internete https://data.kemt.fei.tuke.sk/EVaA_Elektromagneticke_vlny_a_anteny/_materialy/Prednasky/Pr06/Pr06_Uvod%20do%20anten.pdf
20. Pyramidal RF Absorber SA version. Pyramidal RF Absorber SA versionhttp://www.compeng.com.au/document-library/pyramidal-rf-absorber-sa-version/
21. VYŽAROVANIE ELEMENTÁRNYCH ŽIARICOV. Dostupné na internete https://data.kemt.fei.tuke.sk/EVaA_Elektromagneticke_vlny_a_anteny/_materialy/Prednasky/Pr06/Anteny_1cast.pdf
22. Wait, J.R., Hill, D.A.: Theoretical and numerical studies of wire mesh structures. Sens. Simul. Notes, Note 231 (1977)

Chapter 3
Depolarization of a Wave on a Conducting Grid

The first chapter presents the generally known classical non-reflective chambers used in the HF technique. In terms of practical use for various measurements of HF devices, they are fully compliant in terms of dimensions, shape, or power used. The classical non-reflecting chamber has many excellent and desirable properties that allow it to measure even complex and demanding HF circuits and antenna systems. However, the classical non-reflective chamber, through the above significant advantages, has several disadvantages from a practical point of view:

(a) Only large organizations that are able and willing to put considerable funds into its construction can procure such a chamber. The price of an anechoic chamber is derived depending on its physical dimensions and shape and the level of absorption of electromagnetic waves,
(b) prior to such a significant and demanding structure, it is necessary to divide a suitable space or to construct a suitable separate building that can no longer be used for other purposes,
(c) This design is unaffordable in price and space for small production organizations, technically oriented schools, scientific institutions, radio amateurs, and others interested in antenna measurement. Based on the above, the monograph is devoted to designing and analysing a simple but sophisticated system u, which allows the measurement of antennas, even in classical laboratory spaces or standard unmodified rooms. The presented method of measuring the directional characteristics of antennas uses the principle of depolarization over a grid of conductive polarization. This process will be subsequently explained to understand the relatively complex principle of depolarization.

First, it is necessary to familiarize the reader with the fundamental quantities used in describing the depolarization, followed by an explanation of their layout and function in the interaction of the HF signal with the polarization grid. In this step, the most important thing is the knowledge and correct interpretation of the marking of individual vectors of decomposition of the electromagnetic field.

M. Krchňák et al., *Advancements in Antenna Measurement*, Lecture Notes in Electrical Engineering 1108, https://doi.org/10.1007/978-3-031-48835-1_3

Subsequently, the principle of decomposition of the vector of the intensity of the electromagnetic field in its interaction with the material environment, which is a grid formed from straight, parallel-deposited conductors, is demonstrated. At the same time, the conductors of the grid are oriented to one of the three positions, at angles of 0°, 90° and 45°, of the polarization of the transmitted signal. The third step describes the decomposition principle of the HF signal vector when turning the plane of polarization of the incident wave at an angle of 45° to the depolarization panel. This is an essential decomposition of the HF signal vector because it changes the polarization. Thus, in the event of the impact of a vertically polarized wave on the grid, a horizontally polarized wave is reflected. A similar decomposition and change of polarization occur with the impact of a horizontally polarized wave, which is subsequently reflected in the vertical polarization.

3.1 Basic Quantities in the Process of Depolarization

To comprehend the process of depolarization, one must have a grasp of the physical laws governing the decomposition of the intensity vector E of the electromagnetic field; this field is incident on the polarization grid with the RF signal. The RF signal hitting the grid can either ricochet off, pass through, or decompose into other vector components upon contact with the grid. Figure 3.1 illustrates the gradual entry of the RF signal into the depolarization process and its time course.

When explaining the interaction of vector **E** with the grid, we consider the input of three half-waves of the HF signal into the depolarization process. Therefore, in the following section, the three vectors and their components at the different stages of depolarization that will be used in the description are further specified.

1. $\mathbf{E_1}$ vector " + " of the first half of the wave falling on the grid.
2. $\mathbf{E_2}$ vector "–" of the second half of the wave falling on the grid.
3. $\mathbf{E_3}$ vector " + " of the third half-wave falling on the grid.

With the gradual impact of individual vectors (E_1, E_2, E_3) on the grid, the decomposition of every vector into two components occurs:

- $\mathbf{E_{1p}}$; $\mathbf{E_{2p}}$; $\mathbf{E_{3p}}$,a component that passes through the grid.
- $\mathbf{E_{1o}}$; $\mathbf{E_{2o}}$; $\mathbf{E_{3o}}$component that is reflected from the grid.

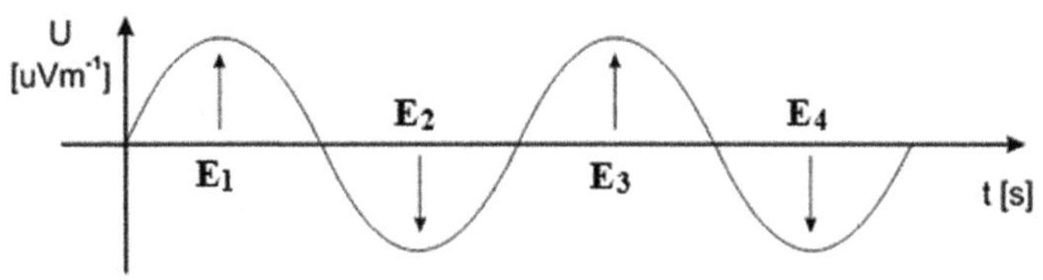

Fig. 3.1 Signal before depolarization—direct with vertical polarization

Individual vectors have the following components after decomposition on the grid:

- $\mathbf{E_{1pas}}$; $\mathbf{E_{2pas}}$; $\mathbf{E_{3pas}}$ vector that passed without turning the phase through the grid.
- $\mathbf{E_{1refl}}$; $\mathbf{E_{2refl}}$; $\mathbf{E_{3refl}}$ vector, embossed from the grid with a 180° phase rotation.
- $\mathbf{E_{1pasrefl}}$; $\mathbf{E_{2passrefl}}$; $\mathbf{E_{3pasrefl}}$ vector, after passing the grid, which is reflected from the conductive plate at a distance of λ/4 behind the grid, rotating its phase 180° and again reaching the level of the grid.
- $\mathbf{E_{1refldly}}$; $\mathbf{E_{2refldly}}$; $\mathbf{E_{3refldly}}$ vector after reflection, after a change in polarization with a delay of λ/2.

3.2 Decomposition of Vector E on a Conductive Polarization Grid

When interacting with a conductive polarization grid made of straight, parallel conductors, the process of depolarization utilizes the decomposition of the electromagnetic field intensity vector E. The wires in the grid are oriented at one of three positions: 0°, 90°, and 45°. Figure 3.2 illustrates the interaction process between Vector E and the conductive grid. This Figure illustrates the element distribution resulting from the interaction between the RF signal and a conductive polarization grid.

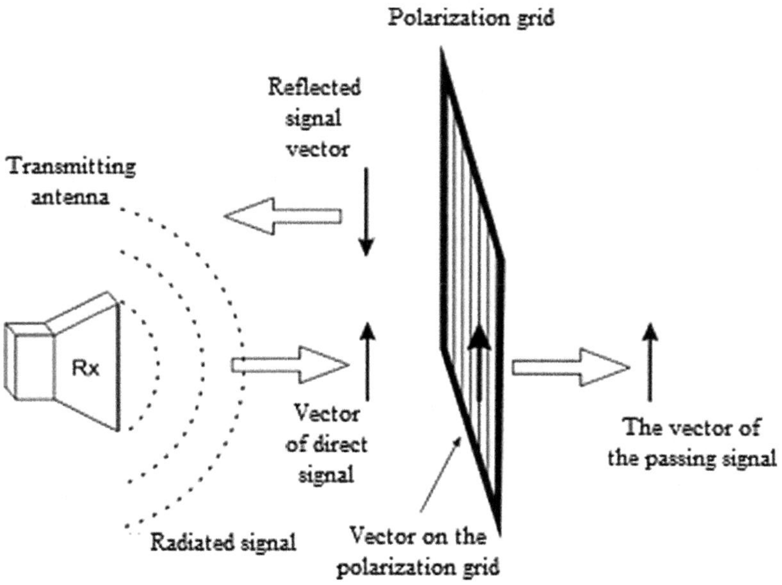

Fig. 3.2 Illustration of the interaction of vector **E** with the grid

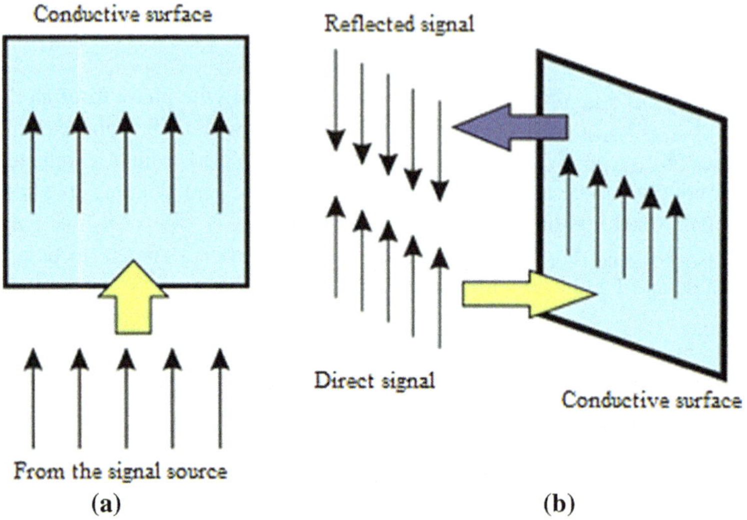

Fig. 3.3 Signal reflection after impact on the conductive surface

The signal originates from an RF generator, and it is transmitted using an antenna. Let's assume that the electric component E's intensity vector, which is the emitted electromagnetic field, features vertical polarization. The direct signal vector in the Figures represents this vector. This vector falls on a conductive polarization grid that consists of straight, parallel-laid conductors. Interference with the vector of the direct signal happens differently depending on the orientation (polarization) of the conductors. Figures 3.4, 3.5, 3.6 and 3.7 demonstrate these three different ways. Figure 3.3 shows comparison of the difference in Vector E's interaction with a conductive polarization grid, in terms of phase ratios after the interaction.

The Figures demonstrate the vector of the direct transmitted signal E with vertical polarization for simplicity's sake. Only the orientation of the conductors of the conductive polarization grid changes. Part (a) of the corresponding Figure highlights the relative position of Vector E of the transmitted signal and the orientation of grid wires. Part (b) of the relevant Figure shows the actual results of the interaction of the relevant relationship.

3.3 Signal Reflection After Impact on a Conductive Plane Surface

Figure 3a depicts the relative positions of Vector $\mathbf{E_p}$ and a regular conductive flat surface (not a grid). We assume the conductive surface plane is at a right angle to the direction of the electromagnetic wave of the transmitted signal, and Vector $\mathbf{E_p}$ lies on the plane when it hits the surface. Figure 3b shows that a complete reflection

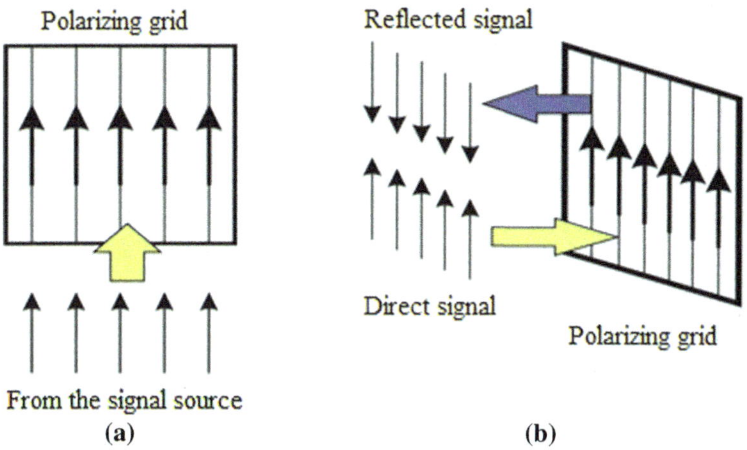

Fig. 3.4 Signal reflection from a vertically polarized grid

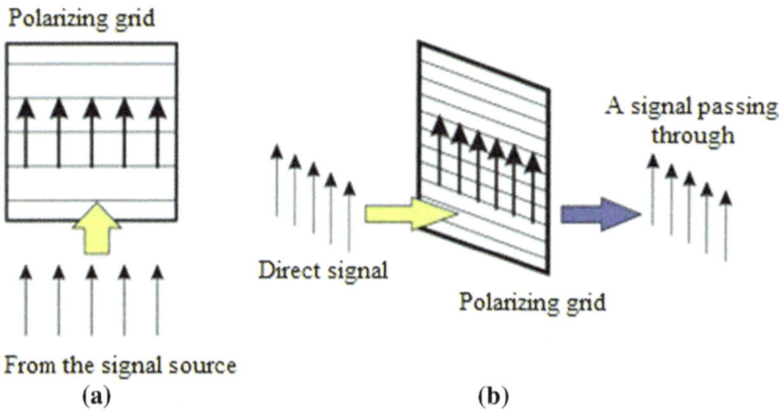

Fig. 3.5 Signal passage through a horizontally polarized grid

of the incident wave occurs upon the interaction between them. Upon reflection, the reflected signal's vector $\mathbf{E_0}$ rotates by 180°. Rotating the ordinary conductive surface around its axis doesn't affect the interaction's outcome, provided the conductive surface's plane remains perpendicular to the direction of propagation of the direct signal. Despite this rotation, the reflected wave is still reflected, meaning that only the phase vector $\mathbf{E_0}$ of the reflected signal experiences a rotation.

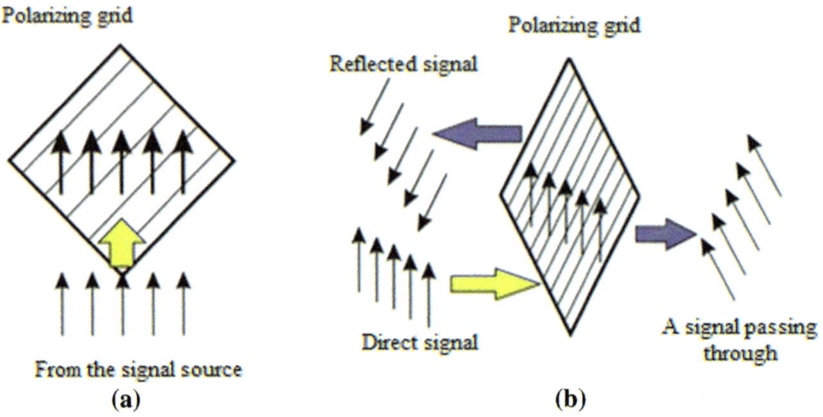

Fig. 3.6 Signal splitting on a grid rotated at an angle of 45°

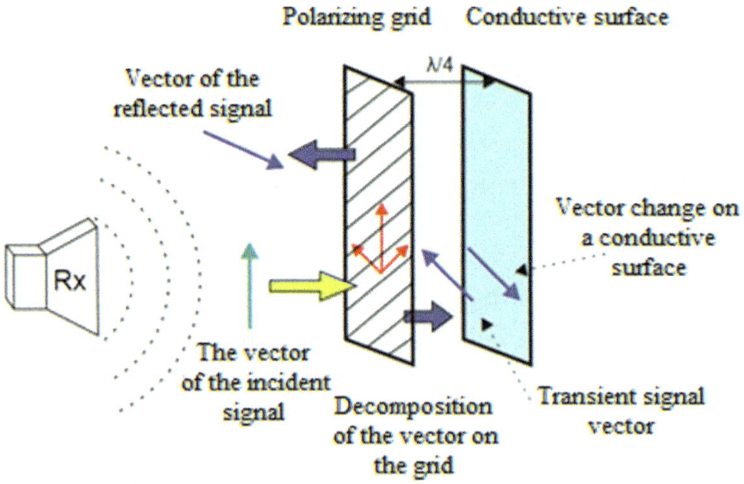

Fig. 3.7 Interaction of the signal with the depolarization panel

3.4 Signal Reflection from the Vertically Polarized Conductive Grid

In Fig. 3.4a an idea of the relative position of the vector $\mathbf{E_p}$ of direct HF signals is shown with the position of the conductors of the polarization grid when the vector $\mathbf{E_p}$ reflects. The vector's and the conductors' orientations are the same, i.e., parallel. In Fig. 3.4b, the reflection of a vertically polarized vector $\mathbf{E_o}$ of reflected HF signal from a conductive grid with vertically oriented conductors is shown. It is assumed that the plane of the conductors is perpendicular to the direction of propagation of the electromagnetic wave of the transmitted signal and that the vector $\mathbf{E_p}$ in the impact, is in this plane. Figure 3.4b shows that during the mutual interaction of vertically oriented vector $\mathbf{E_p}$ of the direct wave with vertically oriented conductors of the grid, there is a complete reflection of the incident wave, in which the vector $\mathbf{E_o}$ of the reflected signal is rotated by 180°. The complete reflection of the $\mathbf{E_o}$ vector occurs only if the distances between the grid conductors are less than $\lambda/4$ of the operating frequency.

Under this condition, at the same time, while maintaining the perpendicular of the planar surface of the polarization grid to the direction of propagation of the direct signal and, at the same time, the $\mathbf{E_p}$ vector orientation of the direct HF signal, with the orientation of the conductors of the polarization grid a complete reflection of the vector $\mathbf{E_o}$ occurs, comparable to the reflection from the planar conductive surface.

If any of the above three conditions are not met, there is no complete reflection of the vector $\mathbf{E_o}$. With a greater distance between the conductors ($h > \lambda/4$), part of the wave passes behind the polarization grid. If the condition of perpendicular of the impact of the vector $\mathbf{E_p}$ to the area of the polarization grid is not met, Snell's law will apply. Suppose the condition of equal vector $\mathbf{E_p}$ orientation with the orientation of the conductors of the polarization grid is not met. In that case, there will be a transition of the wave behind the polarization grid. The same conclusions apply reciprocally to a horizontally polarized vector $\mathbf{E_p}$ of the direct HF signal to the horizontal orientation of the conductors of the polarization grid.

3.5 Signal Passing Through a Horizontally Polarized Conductive Grid

In Fig. 3.5b, the relative position of the vector $\mathbf{E_p}$ of direct wave together with polarization grid conductors' orientation when the vector $\mathbf{E_p}$ is passing is shown. The incident vector is perpendicular to the polarization grid. Figure 3.5b shows the transition of a vertically polarized $\mathbf{E_{pr}}$ vector through a conductive grid with horizontally oriented conductors. It is assumed that the plane of the conductor is perpendicular to the direction of propagation of the electromagnetic wave of the transmitted signal and that the $\mathbf{E_p}$ vector during the impact is located in this plane. While maintaining the perpendicular of the plane in which the conductors lie about the direction of propagation of the direct HF signal, the rotation of the horizontal conductor area of the polarization grid around its axis has a significant effect on the result of the interaction.

In Fig. 3.5b, it is shown that when a vertically oriented vector $\mathbf{E_p}$ of a direct HF signal interacts with horizontally oriented grid conductors, there is a complete penetration of the incident wave through the conductive grid, in which the orientation of the vector $\mathbf{E_{pr}}$ of the passing HF signal does not change. Full vector $\mathbf{E_{pr}}$ penetration through the conductive grid is independent of the distance between the conductors of the grid with respect to the wavelength of the operating frequency. It is also independent of the perpendicular of the planar area of the polarization grid to the direction of propagation of the direct signal. However, for the complete passage of the HF signal through the conductive grid, the perpendicular orientation of the vector $\mathbf{E_p}$ of the direct HF signal to the orientation of the conductors of the polarization grid is required.

If one of the above conditions is not met, there will be no complete vector $\mathbf{E_o}$ transition behind the polarization grid. Reducing the angle below 90° will lower the signal power propagating through the grid. The same conclusions and conditions set out above apply reciprocally to a horizontally polarized vector $\mathbf{E_p}$ of the direct wave in relation to the vertical orientation of the conductors of the polarization grid.

3.6 Decomposition of the Signal on a 45° Polarized Grid

In Fig. 6a the mutual position of the vector $\mathbf{E_p}$ of direct HF signal with the position of the conductors of the polarization grid when the vector $\mathbf{E_p}$ decomposes is shown. There is a 45° angle between the vector's orientation and the conductors' orientation. In Fig. 3.6b, the decomposition of a vertically polarized vector $\mathbf{E_{pr}}$ on a conductive grid is shown. It is assumed that the plane of the conductors is perpendicular to the direction of propagation of the electromagnetic wave of the transmitted signal, and vector $\mathbf{E_{pr}}$ lays in that plane at impact. While maintaining the perpendicular of the plane in which the conductors lie in relation to the direction of propagation of the direct HF signal, the rotation of the area of the oriented conductors of the polarization grid around its axis to an angle other than 45° has a significant effect on the results of the interaction.

In Fig. 3.6b, it is shown that when a vertically orientated vector $\mathbf{E_p}$ of direct HF signal interacts with grid conductors orientated at an angle 45°, there is a decomposition of the vector $\mathbf{E_p}$ of direct wave on this conductive grid. Vector E divides into two components:

- E_o, which is in parallel with the conductors of the polarization grid.
- E_p, which is perpendicular to the conductors of the polarization grid.

Each of these components behaves differently after splitting:

- Parallel component E_o is reflected from the conductive grid with a change of orientation.
- Perpendicular component E_p passes through the conductive grid without changing orientation.

When the conductors of the polarization grid are rotated with respect to the vertically oriented vector $\mathbf{E_p}$ of direct HF signal at an angle of 45°, there is the same energy redistribution at the level of both signals—reflected and passing.

3.7 The Physical Principle of Operation of the Depolarization Panel

When describing the physical principle of operation of changing the polarization of the radio wave on the depolarization panel, all the above-mentioned interactions of the electromagnetic wave with the material environment are used. The concept of the workplace in terms of the distribution of elements in the interaction of the RF signal with the depolarization panel is shown in Fig. 3.7.

Suppose the intensity vector of the electrical component **E** of the radiated electromagnetic field has vertical polarization. The vector of the transmitted signal **E** falls on a conductive polarization grid formed from straight, parallel-mounted conductors oriented to vector **E** at an angle of 45°. Due to the different orientations of vector **E** and the conductors of the polarization grid, interference of the direct signal vector with the grid occurs. When analysing the vector E interference on a depolarization panel, in the process of changing vector polarization (Fig. 3.7), they always consist of vectors from two consecutive half-periods of the signal transmitted with the opposite orientation. We will denote the signal of the first half of the period by the index $\mathbf{E}_1$, the signal of the second half of the period by the index $\mathbf{E}_2$, the signal of the third half of the period by the index $\mathbf{E}_3$, etc., according to Fig. 3.1. At the impact of the direct signal vector $\mathbf{E}_1$ of the first half of the period to a conductive grid at an angle of 45° Fig. 3.6a, it first decomposes into two new, perpendicular vector components (Fig. 3.8b), similar like on Fig. 3.6.

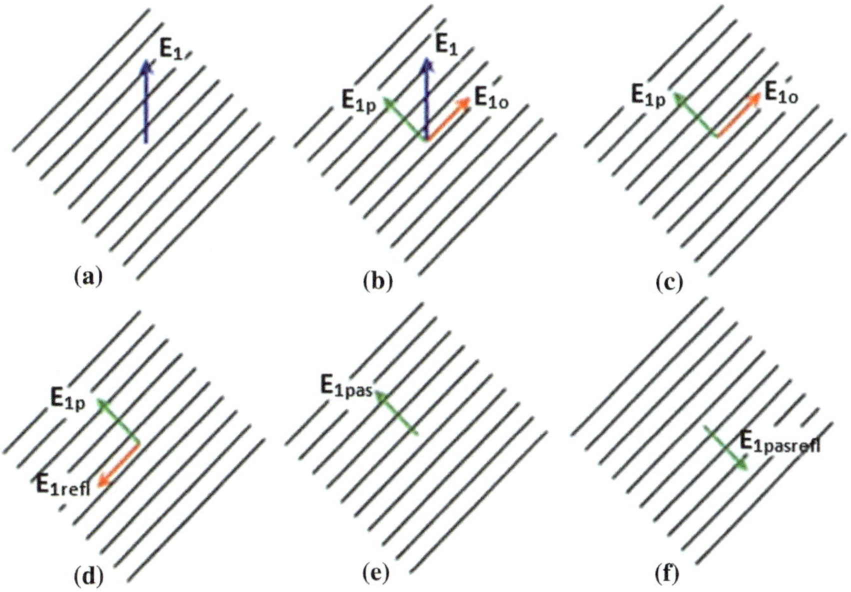

Fig. 3.8 Decomposition of the E_1 vector with depolarization panel grid

Each of these two components (Fig. 3.8c) further interferes with the conductors separately. The first component $\mathbf{E_{1o}}$ Fig. 3.8c is oriented parallel to the conductors of the polarization grid and thus immediately reflects from them after impact with the opposite 180° rotated phase as a component $\mathbf{E_{1refl}}$, (Fig. 3.8d). The second component $\mathbf{E_{1p}}$ (Fig. 3.8c), perpendicular to the conductors of the polarization grid, passes through this grid without attenuation and with the same phase as the component $\mathbf{E_{1pr}}$ (Fig. 3.8d). This component $\mathbf{E_{1pas}}$ continues further thru the grid separately (Fig. 3.8e).

Behind the conductive polarization grid, at a distance of λ/4, the conductive surface of Fig. 3.7 is located. In this area, the decomposition of the vector occurs E_{1pas}, which has passed through a polarization grid, similarly, as described in Fig. 3.3. This means that it will be reflected from the planar conductive surface as a vector $E_{1pasrefl}$, but with a rotated orientation of 180° (Fig. 3.8f). Reflected vector $E_{1pasrefl}$ reappears at the level of the depolarization grid, but this time from behind. When overcoming the distance from the grid to the planar surface and back (2λ/4), a time proportional to λ/2 elapses. It is at the moment when the vector of the second half of the period falls on the polarization panel from the front E_2.

At the impact of the vector $\mathbf{E_2}$ of the transmitted signal of the second half of the period on the conductive grid at an angle of 45° (Fig. 3.9a), it also splits into two new, each other perpendicular vector components, (Fig. 3.9b). It is similar to (Fig. 3.8) but only with the opposite orientation. These two components (Fig. 3.9c) further interfere with the conductors separately. The first component $\mathbf{E_{2o}}$ (Fig. 3.9c), is oriented in parallel with the conductors of the polarization grid and thus immediately reflects from them after impact with the opposite 180° rotated phase as a component $\mathbf{E_{2refl}}$ (Fig. 3.9d).

Reflection of the component E_{2refl} from the polarization grid, is carried out at the moment when the vector falls on this grid from during the rear $E_{1pasrefl}$ the first half of the period from the plane surface. Thus, the reflected component of the vector (from the planar surface—the first half of the period) $E_{1pasrefl}$ (Fig. 3.10a) and the reflected component of the vector (from the polarization grid—the second half of the period) E_{2refl} (Fig. 3.10b) meet at the level of the plane of the grid at the same time, Fig. 3.10c. This fact will allow them to be reassembled into one resulting vector, E_{1refl}, Fig. 3.10d. The resulting reflected vector E_{1refl}, Fig. 3.10e returns from the depolarization panel, based on the rules of Snell's law back.

The second component E_{2p}, Fig. 3.9c perpendicular to the conductors of the polarization grid passes through this grid without attenuation and with the same phase as the component E_{2pas}, Fig. 3.9d. Ingredient E_{2pas} continues further behind the grid separately, Fig. 3.9e. From the conductive plane surface located behind the polarization grid at a distance λ/4, the vector component is reflected with the phase rotated 180° against the vector $E_{2pasrefl}$, Fig. 3.9f. Reflected vector $E_{2pasrefl}$ appears from the rear again at the level of the depolarization grid at the moment, when the vector of the third half of the period falls on the polarization panel from the front E_3. At the impact of the direct signal vector of the third half of the period E_3 on a conductive grid at an angle of 45°, it also occurs due to its decomposition into two new, perpendicular vector components to each other, similar to Fig. 3.8. First

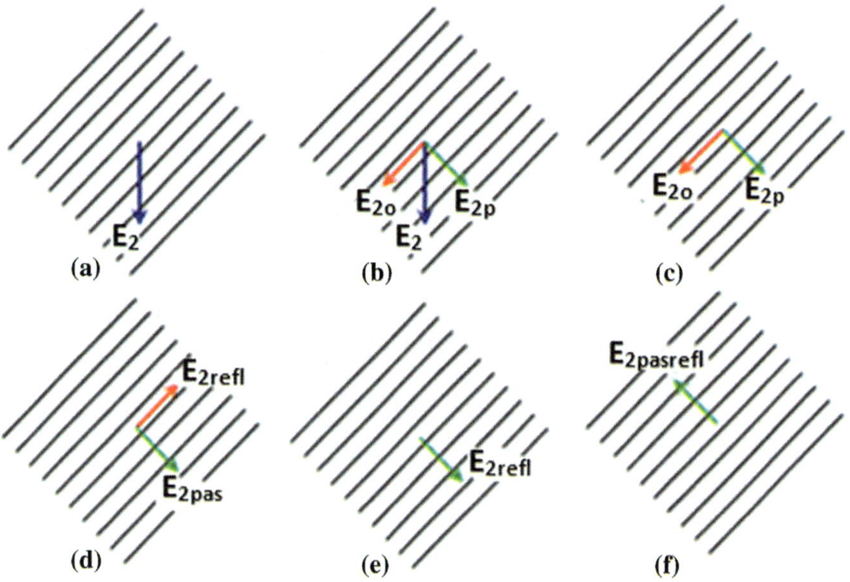

Fig. 3.9 Splitting of the vector E_2 when falling on the depolarization panel

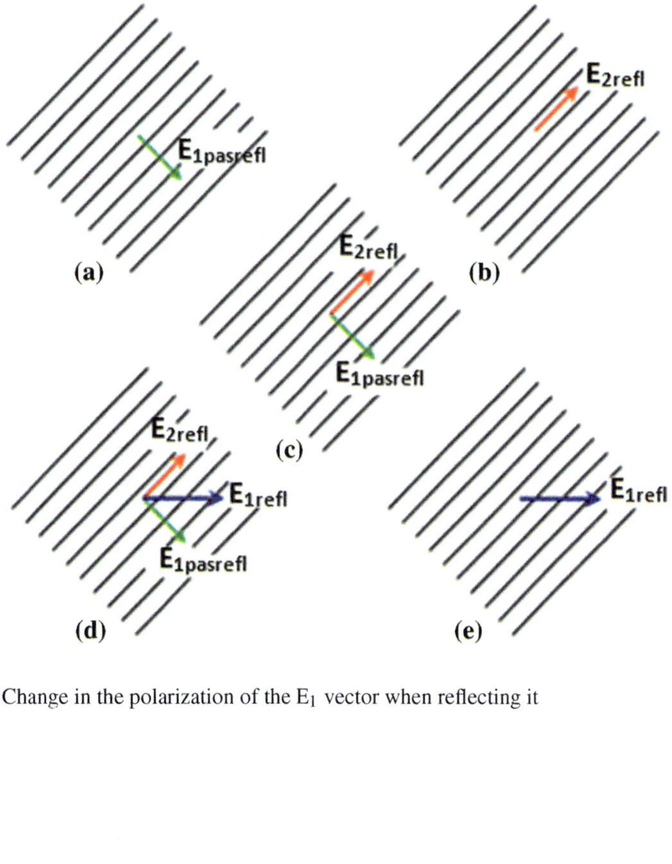

Fig. 3.10 Change in the polarization of the E_1 vector when reflecting it

component E_{3o}, there is oriented parallel to the conductors of the polarization grid, i.e., after impact it is immediately reflected with the opposite, 180° rotated phase. Reflection of the ingredient E_{3refl}, Fig. 3.11b occurs on the polarization grid at the moment when the vector falls on this grid from behind $E_{2pasrefl}$ during the second half of the period from the plane surface. Thus, the reflected component of the vector (from the planar surface—the second half of the period) $E_{2pasrefl}$, Fig. 3.11a, and the reflected component of the vector (from the polarization grid—the third half of the period) E_{3refl}, Fig. 11b, meet at the level of the plane of the grid at the same time, Fig. 11c. This will allow them to be reassembled into one resulting vector E_{1refl}, Fig. 10d. The resulting reflected vector E_{1refl}, Fig. 3.10e, returns from the depolarization panel based on the rules of Snell's law back.

As a result of the process described above, the depolarization panel itself occurs:

(a) a 45° change in the polarization of the HF field intensity vector,
(b) to change the phase of the HF signal by 180°,
(c) the time delay of the HF signal by λ/2 (Fig. 3.12).

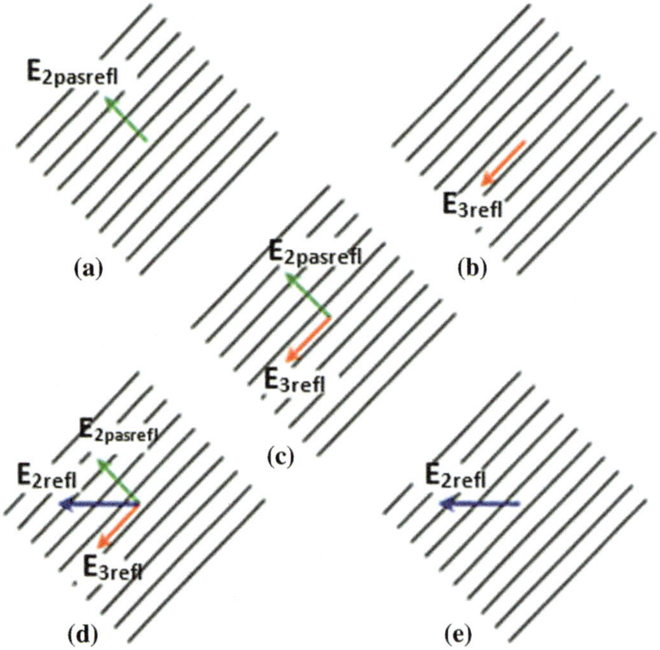

Fig. 3.11 Change in the polarization of the E_2 panel vector when reflecting it

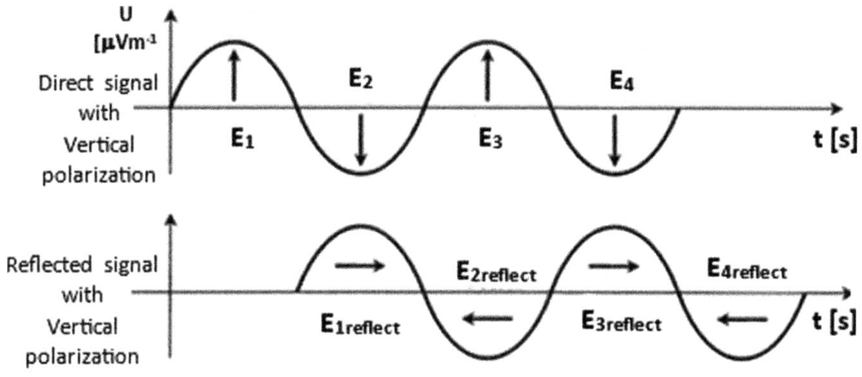

Fig. 3.12 Time shift of vector E in the process of depolarization

3.8 Simulation of a Nearby Field Around the Grille Wires

The previous theoretical analysis explained the behaviour of the electromagnetic wave on the polarization grid. To complement these theoretical assumptions, a simulation of the near field around a parallel array of conductors in free space was developed. The simulation was made in a simplified form, simulating a near field around conductors excited by a homogeneous electromagnetic field with linear polarization. According to the available information, the CADFeko simulation program does not allow the simulation of phase relationships [1]. The results of the simulation contribute to the subsequent optimization of the panel design. Due to the limitations of the available computer technology, the simulation had to be performed on a reduced scale.

The thickness of the simulated conductors is $\varphi = 1.6$ mm. There is a gap of 1 mm between adjacent wires. The total number of drivers is 40. The length of the wires is $l = 50$ mm. Wires of this length will be in full-wave resonance at a frequency of 6 GHz. Considering that the wires will also be in resonance for $\lambda/4$ and $\lambda/2$, representing frequencies of 1.5 and 3 GHz, we arrive at the working range of the depolarization field. Figure 3.13 shows a preview of the simulated structure of the conductors shown in orange. The wires are symmetrical to the X and Y axes.

The red-blue arrow pointing to the centre of the system represents the direction of the electromagnetic wave and the orientation of the vectors of the electric and magnetic components. The simulation has been made for both horizontal and vertical polarization and wave polarization. Due to the symmetrical structure of the simulation object, it was possible to simplify the calculation by setting the symmetry axis for the X and Z planes. The blue part shows the environment of the object and the area in which the near field is investigated. This area has dimensions in the X axis = 140 mm, in the Y axis = 20 mm and in the Z axis = 80 mm. The simulation of the field is carried out after certain steps in each axis.

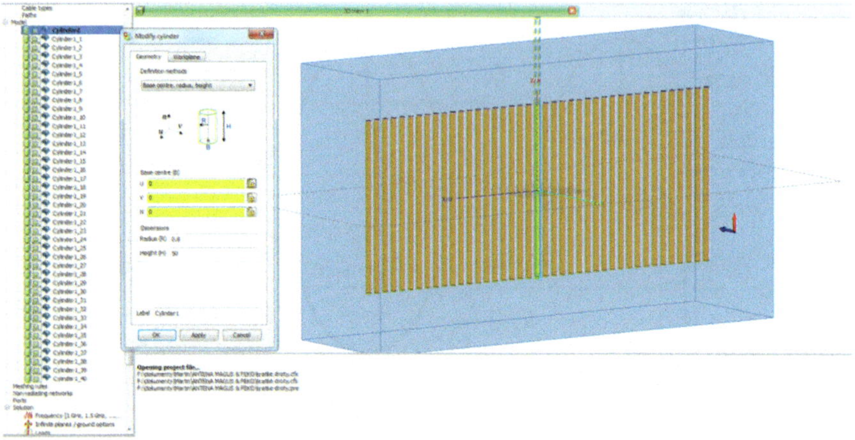

Fig. 3.13 Simulation of the structure of conductors on the grid

There are 9 steps for each axis. The results of the simulation can be divided into two basic planes in which they are displayed:

- YZ plane, showing the field strength by looking at the wires from left/right;
- XZ plane, showing the field strength by looking at the wires from the front/rear;

The results show the electric and magnetic components of the field. The results are displayed using a colour scale, where blue shows the minimum field strength and red shows the maximum strength [1].

Since the evaluation of the simulations occurred only in the form of monitoring the field distribution, i.e., the areas where the maximum and minimum radiation occurs and how the wave passes through the conductors, the goal was not to evaluate the absolute levels of the electrical and magnetic component of the field on the conductors. Therefore, there was no need for these scales (Fig. 3.14) to indicate each result.

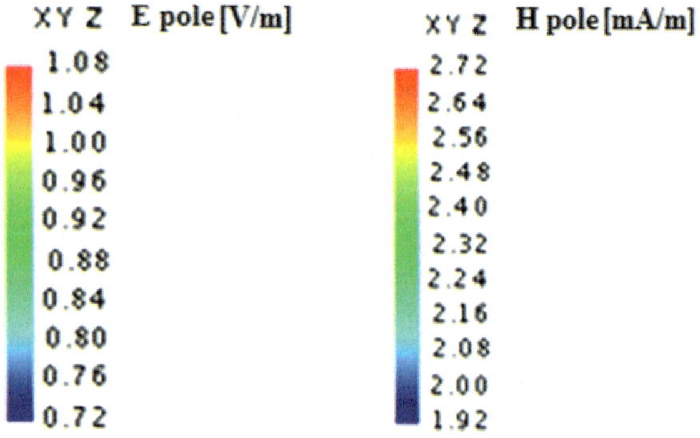

Fig. 3.14 Example of a colour scale representing the results of the simulation

3.9 Simulation of a Near Field with Parallel Wave Polarization

In the first simulation, the polarization of the wave was set parallel to the direction of the conductors (the vector of the electric component is parallel to the direction of the conductors). Such a state does not occur with the real use of the depolarization panel. However, the results complement the theoretical background and help to understand how an electromagnetic wave behaves with the same polarization as the orientation of the grid conductors.

By observing the field at a distance of 0 mm from the centre of the wires, we get an illustration of the field in the centre of the driver arrangement. By analysing Fig. 3.15, it was found that the magnetic field inside the grid is almost immutable with increasing frequency. In the field, areas with a slight (zero) magnitude of the magnetic field alternate with areas with a certain level of magnetic field. The grille of the conductors appears to be in a state of resonance. The state of the resonance varies according to the ratio l/λ. This claim is confirmed by the fact that at the frequencies 1.5 and 3 GHz the field is identical. At these frequencies, the grid is in λ/4 and λ/2 resonance.

By changing the distance of the monitored field to 5 mm, we get to the area behind the wires. At a frequency of 1 GHz, from Fig. 3.16, the maximum of the magnetic field is at the edges of the conductors, around the grid, there is a magnetic field close to 0. By increasing the frequency, the wires enter the area of resonance. In all cases, the maximum magnitude of the magnetic field at the edges of the conductor is visible. The magnetic field around the conductors changes only slightly for the cases of 2, 3, 2.5 and 4 GHz. At frequencies of 1, 1.5, and 2.5 GHz, the magnetic field around

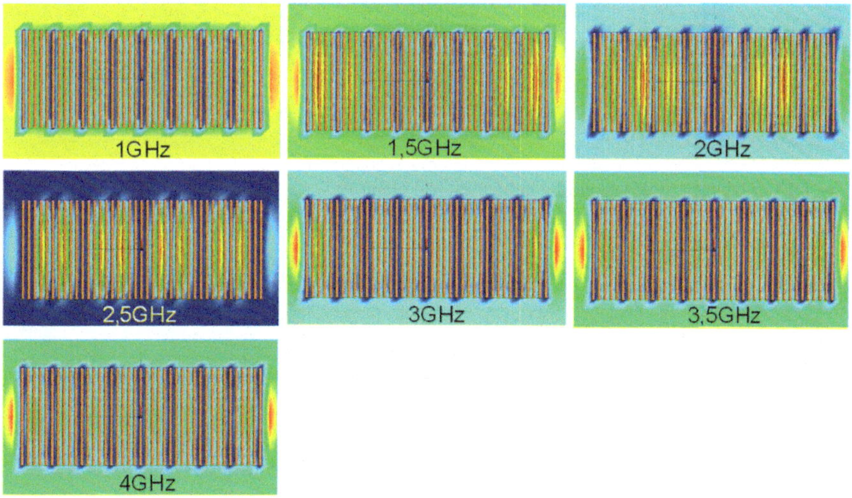

Fig. 3.15 Magnetic field distribution, XZ plane

the conductor is not uniform, but changes towards the centre of the conductor. For a frequency of 2.5 GHz, the maximum magnitude of the field is in the middle of the grid. In other cases, it is on the edge of the driver (Fig. 3.17).

Electric fields (Fig. 3.18).

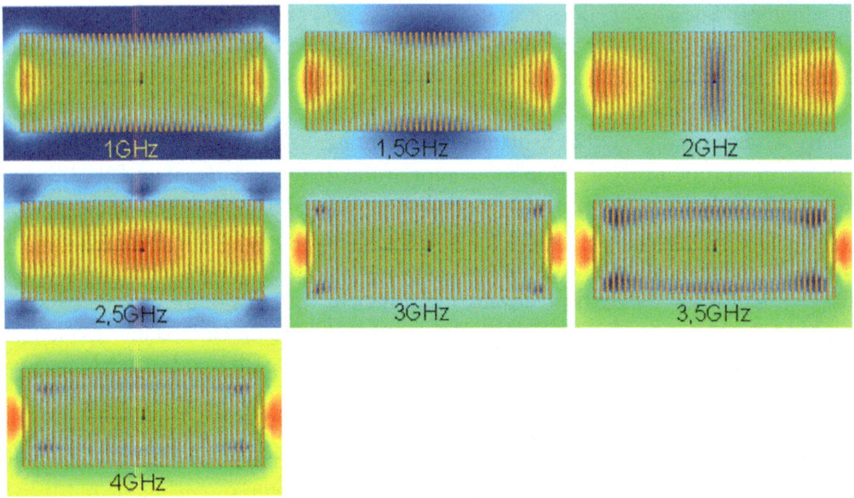

Fig. 3.16 Magnetic field distribution, XZ plane

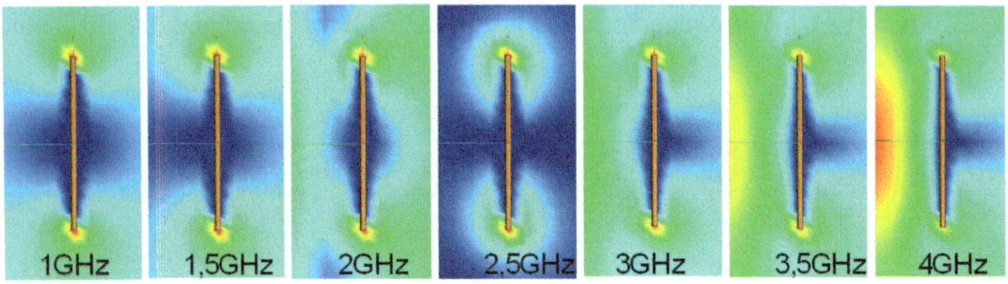

Fig. 3.17 Magnetic fields around conductors, YZ plane

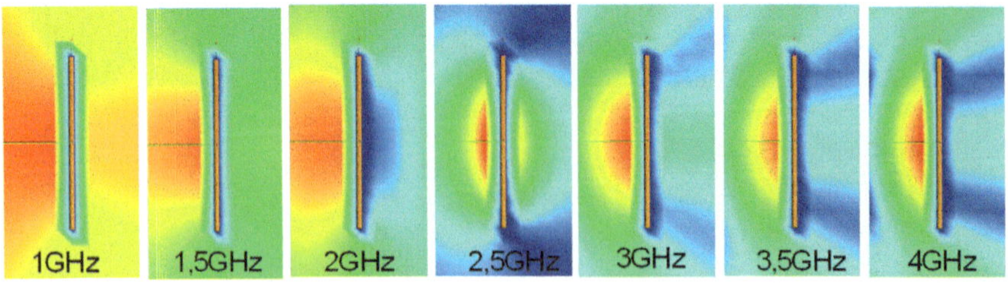

Fig. 3.18 Electric fields around conductors, YZ plane

The view of the YZ plane when watching an electric field shows how, at a 1 GHz frequency (the wavelength is greater than the length of the conductor), the conductor's field flows without significant influence. By reducing the wavelength, the wave is influenced by the conductors to the point that only a minimal part of the wave passes behind the conductor grid. This influence is noticeable at frequencies outside 1 GHz. Maximum magnitude field forms in front of the conductors. At a frequency of 2 GHz, there is a field of almost zero intensity behind the conductors. At a frequency of 3 GHz, a maximum area is formed in front of the conductor, but behind the conductors the field level is minimal.

In Fig. 3.19, the electric field inside the wires is displayed. The magnitude of the electric field is close to zero at given frequencies inside the conductor. This confirms the theoretical assumption that a wave parallel to the parallel arrangement of conductors is reflected from the conductors and only a minimal part of the wave passes through these conductors.

Analysis Fig. 3.20 provides an essential insight into the future construction of the depolarization panel. A near-field simulation demonstrates that the maximum electric field is at the end/edge of the conductors. At a frequency of 1 GHz, the wavelength relative to the conductor is large, the field only partially affects the conductor, and the maximum field strength is concentrated on the edges of the conductor. This fact is also supported by Fig. 3.16, where at a frequency of 1 GHz, the wave can be seen that the wave is circling the conductor and the interaction with the conductor is minimal. With a reduction in the wavelength, the conductors begin to react to the electromagnetic wave at an increased rate. The wave "hits" the wires, and with increasing frequency, stops circling the wires and is blocked by the wires. The conductors are excited by this field and there is field radiation at the edges of the drivers, resulting in the

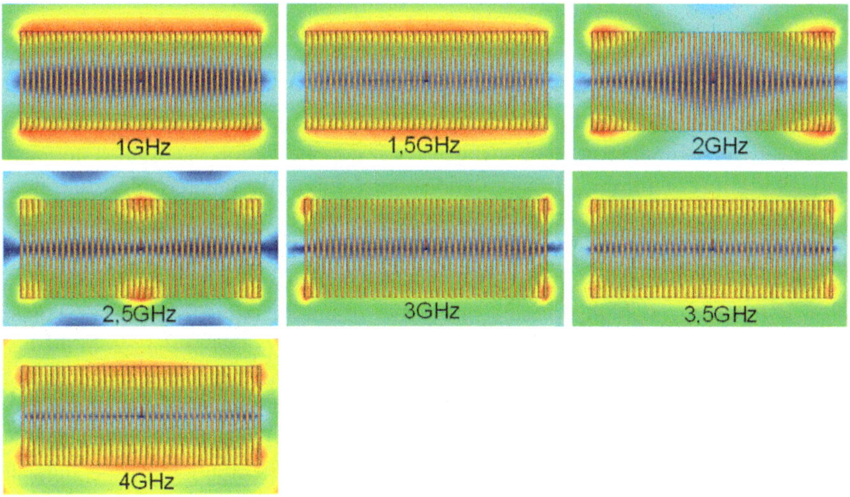

Fig. 3.19 Electric field distribution, XZ plane

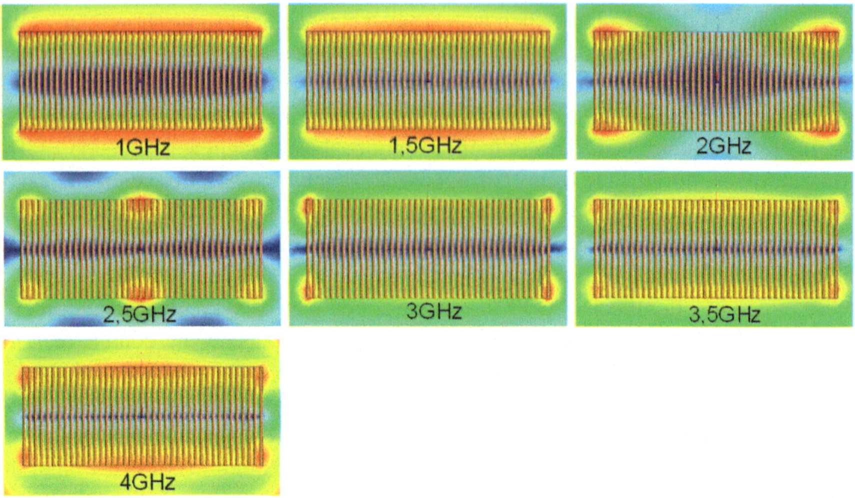

Fig. 3.20 Electric field distribution, XZ plane

formation of a secondary field behind the conductors, which ideally should not be there. The theoretical assumption in Chapter 3 does not describe this secondary field.

3.9.1 Simulation of a Near-Field with Perpendicular Wave Polarization

The simulation has been modified so that the vector of the electric component is perpendicular to the direction of the conductors. In the case of real measurements, the state when the transmitting antenna is vertically polarized and the experimental depolarization panel is at an angle of rotation of 0° (the wires are horizontal). As in the previous simulation, such a state on the depolarization panel does not occur under working conditions. The simulation complements the theoretical assumptions.

In this Chapter describes the decomposition of an electromagnetic wave on a grid of conductors if the conductors are perpendicular to the polarization of the electromagnetic wave. The theory defines that in such an arrangement, an electromagnetic wave will pass through the grid of conductors without the influence of the grid on the wave. This claim is confirmed by the simulations in Figs. 3.21 and 3.22. In the case of a tracking field at a distance of 0 mm from the conductor, the wave passes through the conductor, which is shown in red in the conductor area. There is a minimum level field around the drivers (Fig. 3.23).

By moving the boundary at which the level and distribution of the field are observed in the simulation to 5 mm behind the wires, it is shown in colour. The wave passes through the wires at all frequencies and is not significantly affected by these conductors (Figs. 3.24).

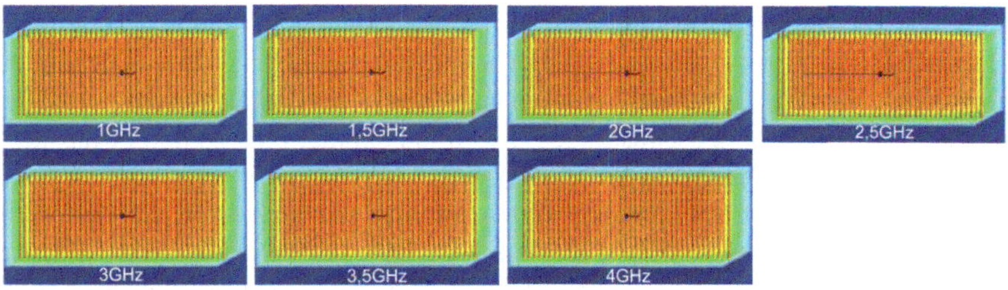

Fig. 3.21 Electric field distribution, XZ plane

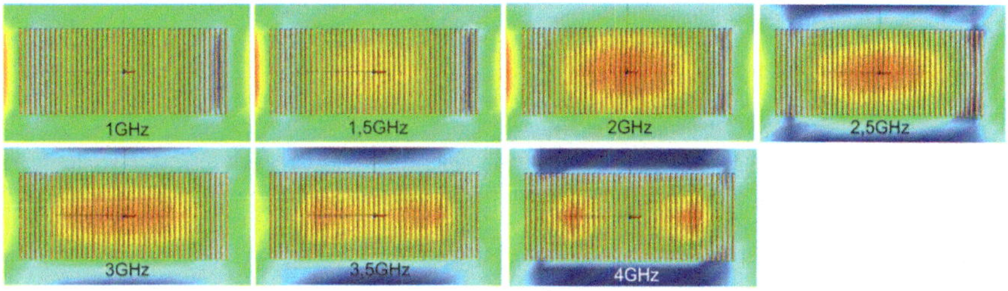

Fig. 3.22 Electric field distribution, XZ plane

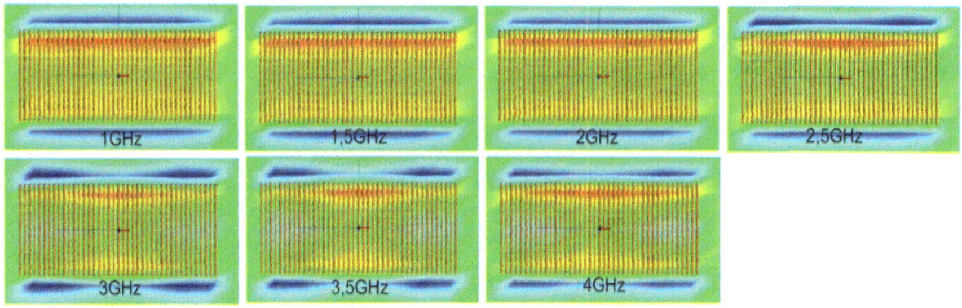

Fig. 3.23 Magnetic field distribution, XZ plane

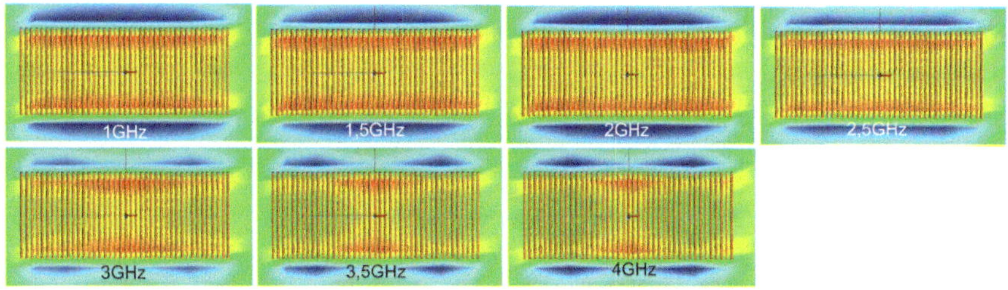

Fig. 3.24 Magnetic field distribution, XZ plane

In the case of magnetic field monitoring, the situation is similar to that of an electric field. In all observed cases (0 mm behind the conductor and 5 mm behind the conductor), the field layout is the same. The effects of the conductors are not visible on the field distribution, as was the case with the parallel polarization of the wave with the conductors in passing subheading 2.8.1. At resonant frequencies of 1.5 and 3 GHz, no changes at other frequencies are visible, which confirms the theoretical assumption that the wave passes through the conductors without their influence. Therefore, it is not necessary to analyse the distribution of the electromagnetic field in the plane YZ.

Reference

1. Wiles, M., Rodriguez, V.: Choosing the right chamber for your test requirements, ETS-lindgren. Dostupné na internete (2010). https://interferencetechnology.com/choosing-the-right-chamber-for-your-test-requirements/

Chapter 4
Standard Methods for Analyzing Antenna Parameters

This chapter describes methods for measuring selected antenna parameters. These are mainly methods for measuring impedance parameters, measuring antenna radiation patterns and measuring gain. Measurements in antenna technology require a wide range of different measurement methods and instrumentation. Typically, these range from simple measurements such as impedance matching and transmission measurements, to high-resolution radiation pattern measurements and long-range pulse response measurements. Many antenna measurements require shielded anechoic chambers, reflectors, test antennas with known parameters, handling equipment and cost-effective software.

4.1 Impedance Parameters in Antenna Technology

The measurement of the S-parameters (S_{11}, S_{12}, S_{21}, S_{22}) of two antennas shall be considered a two-port network measurement. These parameters are measured using a network vector analyzer. Vector analyzers are the most versatile and complex instruments in RF technology. The instrument is capable of characterizing RF elements by measuring their network parameters—with parameters as a function of frequency, to also detect noise and non-linear characteristics. The vector analyzer can measure many parameters in a single measurement, providing an almost complete characterization of the connected circuit. The S-parameters describe the magnitude and phase relationships between the incident and reflected waves. They are numbered according to where the wave begins and where it propagates. The term S-parameters is derived from the scattering matrices described in the 1965 IEEE article "Power Waves and the Scattering Matrix" [1]. The number of S-parameters is given by the square of the number of DUT ports. The two-port device has four S-parameters. The numerical designation of the parameter indicates where and how the wave travels.

M. Krchňák et al., *Advancements in Antenna Measurement*, Lecture Notes in Electrical Engineering 1108, https://doi.org/10.1007/978-3-031-48835-1_4

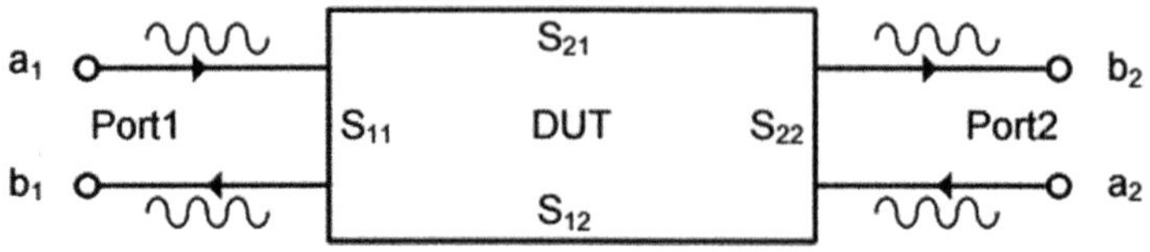

Fig. 4.1 Two-port model S parameters [5]

The first number is the port from which the wave is emitted, and the second number is the port into which the wave enters.

For example, parameter S21 represents a measurement where the signal is generated by port 2 and port 1 is excited by this RF signal after passing through the DUT. If the numbering is identical, it is a reflection coefficient measurement. The forward S-parameters are determined by measuring the magnitude and phase of the incident, reflected and transmitted signals, and the port is loaded with an impedance identical to that of the test system. In the case of a two-port DUT, S11 is the complex reflection coefficient or impedance of the DUT and S21 is the complex transmission coefficient. By connecting the signal source to the DUT output port and loading the DUT input with the ideal load, it is possible to measure the return parameters. The parameter S22 is the equivalent of the output complex reflection coefficient or output impedance and S12 is the reverse complex transmission coefficient[2, 3, 4] (Fig. 4.1).

Several significant models can be found in electronics, e.g., impedance models described by Z parameters, admittance models described by Y parameters, and hybrid models. These are described by H parameters, and from them are derived S parameters [1, 2].

The basic equations describing the two-port S parameters are:

$$b_1 = S_{11}a_1 + S_{12}a_2 \tag{4.1}$$

$$b_2 = S_{21}a_1 + S_{22}a_2 \tag{4.2}$$

It follows from Eqs. (4.2) and (4.3) that the wave leaving DUT ($b_1|b_2$) is a linear combination of the wave entering the DUT ($a_1|a_2$). Loading the DUT with the characteristic impedance Z_0 will allow the calculation of all S parameters. For example, if the characteristic impedance Z_0 is equal to 50 Ω, and if the load impedance of port is 50 Ω, ports 2, and 2 acquire zero values, thus getting the equations for S_{11} and S_{21} [2, 5].

Parameter S_{11} is the reflection coefficient at the DUT input. It is defined as the ratio of the reflected and incident wave at the input.

$$S_{11} = \frac{b_1}{a_1}|_{a_2=0} \tag{4.3}$$

Parameter S_{21} characterizes the transmission of the circuit in a straight line.

$$S_{21} = \frac{b_2}{a_1}|_{a_2=0} \tag{4.4}$$

The absolute value of this parameter can reach any value. The positive or negative value then determines whether the circuit has gain or attenuation.

Parameter S_{22} is a return parameter and determines the coefficient of reflection at the output of the DUT circuit.

$$S_{22} = \frac{b_2}{a_2}|_{a_1=0} \tag{4.5}$$

By analogy, the parameter S_{12} is a reverse transmission [2].

$$S_{12} = \frac{b_1}{a_2}|_{a_1=0} \tag{4.6}$$

S parameters are vectors, so they carry information about magnitude and phase as a function of frequency. The measured parameters can be easily used to link measurements to simulation, as they can be stored in a standardized format, e.g., Touchstone [2].

4.2 Application of a Vector Analyzer in Antenna Technology

At high frequencies, it is virtually impossible to measure with an oscilloscope or conventional voltmeters. In the case of antenna elements or printed RF elements, the signal magnitudes are in the order of uV to mV and the frequencies in the tens of MHz to tens of GHz. In this case, analysis of the entire circuit (network) is used. The VNA—Vector Network Analyzer is used for these analyses. With its help, it is possible to analyse simple devices such as filters and amplifiers as well as very complex and sophisticated modules for satellites and others. Vector analysis are the most versatile and complex instruments in RF technology. They are capable of characterizing RF elements by measuring their network parameters—S-parameters as a function of frequency—as well as detecting noise and non-linear characteristics, including intermodulation and compression. As a result, the vector analyzer can measure many parameters in a single measurement, providing an almost complete characterization of the connected circuit. A simpler form of network analyzer is the scalar analyzer. This measures only the differences between the amplitudes of the signals (Fig. 4.2).

The basic principle of a vector analyzer is the measurement of impedance. Measuring impedance at low frequencies is simple, it is enough to use a signal generator, a voltmeter and an ammeter and calculate the impedance based on Ohm's law. For high frequencies and microwaves, the measurement is much more complex.

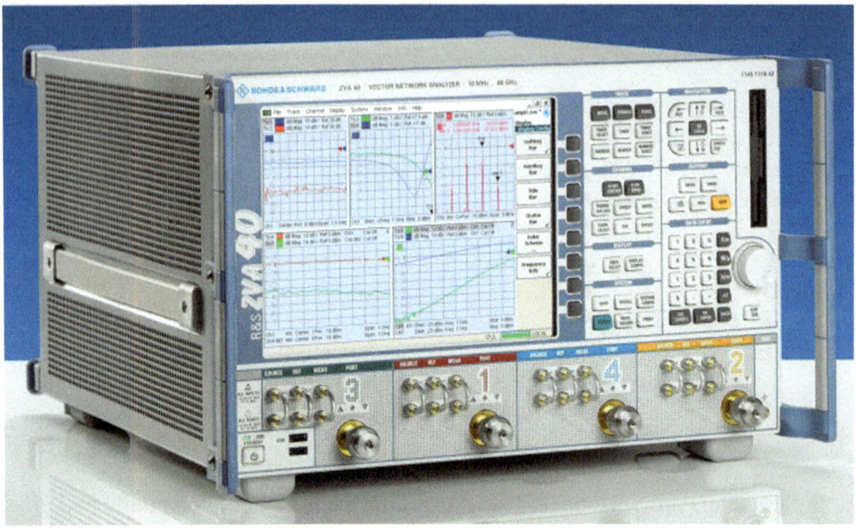

Fig. 4.2 Four-channel vector analyzer

Analyzers use the measurement of incident and reflected waves. It is based on the principle of measuring impedance at low frequencies but adapted to RF. The impedance or reflection coefficient analyzer stimulates the DUT (device under test) with a sinusoidal signal generated by an internal generator. The voltmeter and ammeter are replaced by two receivers. These receivers, together with isolation circuits, evaluate the response of the DUT by measuring the phase and amplitude of the incident and reflected waves from the DUT. The analyzer must have a calibration facility to eliminate systematic errors and calculate the appropriate ratios required to determine the S-parameters. To determine the general characteristics of an unknown linear device, it is necessary to make measurements under different conditions and calculate a series of parameters. These parameters can then be used to determine the overall behaviour of the circuit. The vector analyzer can completely suppress the system measurement error. The magnitude of the S-parameter corresponds to the ratio of the amplitudes of the waves. The phase S parameter is the phase difference between the waves. The reflected wave is measured by channel A and the emitted wave is measured by channel B. The data measured by the vector analyzer can be converted to the time domain, which provides more possibilities for further data processing.

Description Fig. 4.3 (1) magnitude S21; Fig. 4.3 (2) Smith diagram S11(R + jx); Fig. 4.3 (3) phase S21; Fig. 4.3 (4) standing wave ratio SWR.

By measuring and calculating the vector reflection coefficient, it is possible to construct a Smith diagram. Figure 4.3 shows some basic transmission parameters. The transmission coefficient is defined as the ratio of the reflected wave to the emitted wave. If the absolute value of the reflected wave is greater than or equal to the emitted wave, the DUT has a gain. If, on the other hand, this value is less, the DUT

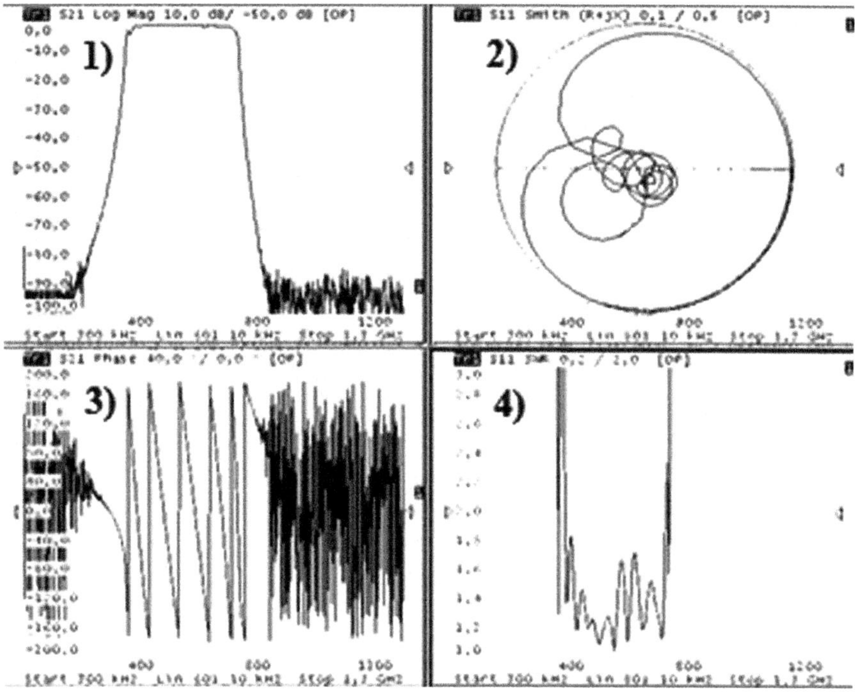

Fig. 4.3 Example of measurements from a vector analyzer

has transient losses. There are two basic principles of signal processing in vector analyzers. The first uses the principles of a direct mixing receiver. It has an oscillator that provides a signal at the output and, at the same time, a signal for processing the received signal. This type of analyzer is simpler and more economical. Because of their simplicity and technical limitations, they provide only simple measurements. The second type uses the superheterodyne principle and provides more accurate measurements. Their complexity affects the price, but they can operate over a wide range of frequencies (up to hundreds of GHz) (Fig. 4.4).

The block diagram has four basic parts. The input circuits separate the incident and reflected waves at the input terminals. The signal is then fed to either the measurement or reference channel for processing. Electronic attenuators are used to adjust the output level of the signal. The generator provides an RF signal that excites the input DUT. This signal is fed through the switch to the selected test port. This port is then marked as the active test port. Each of the input circuits is combined with two independent receivers—a reference and a test receiver of the superheterodyne type. The output from the superheterodyne continues to the digital section, where there are several aliasing filters, precision analogue-to-digital converters and a DSP processor. At the output of this channel is a set of raw data in digital form, ready for further processing. It goes to a powerful computer. The computer provides the graphical

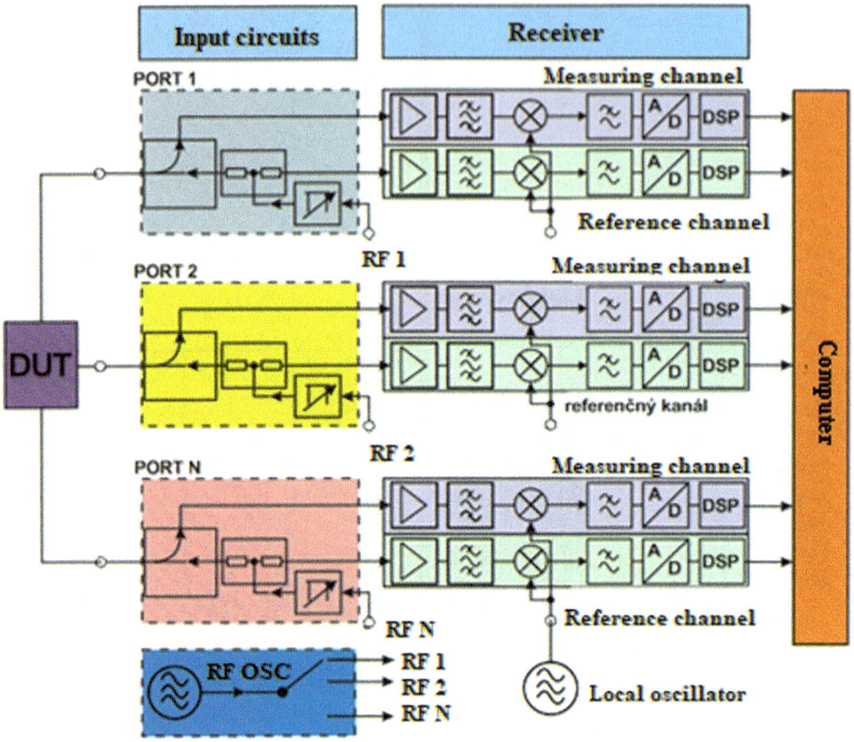

Fig. 4.4 Block diagram of N port analyzer [6]

interface, and all the necessary calculations, and performs system error correction [7].

4.3 Calibration of the Network Analyzer for Analysis of S Parameters

The basic general scheme of a 4-channel network analyzer is shown in Fig. 4.5. It is based on a high-frequency oscillator connected to the output terminals of the analyzer via a switch. The switch allows the oscillator to be switched to the PORT 1 or PORT 2 terminals to which the DUT is connected. It also provides a load on the input port with a load impedance of Z0. The RF signal is picked up by directional taps. They detect and separate the incident and reflected waves from the transmitted signal in both directions (forward and backwards). The outputs of the directional taps are fed to the inputs of the IF mixers which produce a constant low frequency IF signal. The local oscillator is tuned to the frequency VF + MF [6, 8].

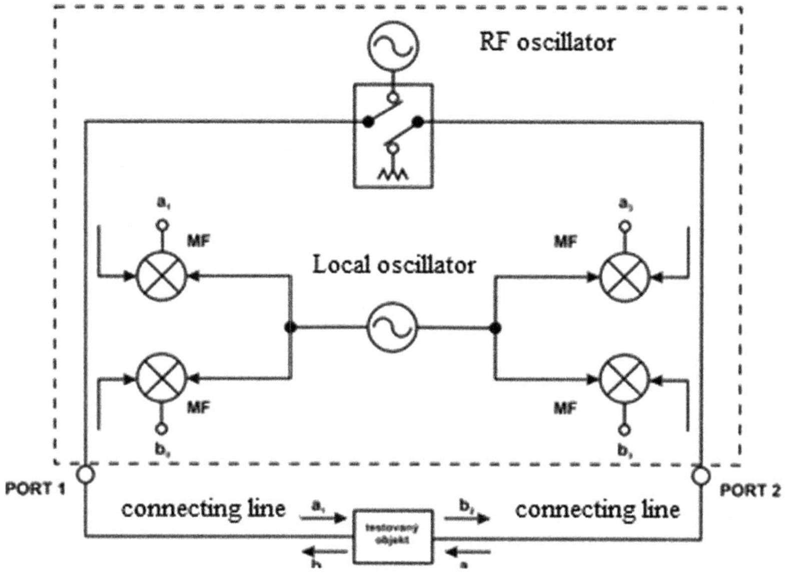

Fig. 4.5 General scheme of a 4-channel network analyzer [6]

From the general scheme it is clear that the system contains a number of complex components which are to some extent the source of measurement errors. The vector analyzer is an almost perfectly linear device over a wide frequency range and has a very clean spectrum of the transmitted signal. To achieve such characteristics, it is necessary to calibrate the instrument repeatedly.

This eliminates errors caused, for example, by impedance mismatch of connectors or signal attenuation on the cabling. At the same time, the calibration process removes any residual signal, even if a high-quality load is connected to the device's terminals [9, 6].

There are a number of calibration algorithms, some of which are built into the instrument and performed automatically. Calibrations that require the use of calibration kits prior to measurement must be selected on the basis of measurement accuracy requirements. Calibration selection consists of two basic steps. The first is to select the port on which the calibration is to be performed. The second step is to select the appropriate calibration algorithm, i.e., how the calibration will be performed [9, 6] (Table 4.1).

Calibration algorithms are defined by various abbreviations that are inconsistent in the literature. The letters in the algorithm label often have different meanings, depending on how the device manufacturer defines them. Among the most basic are algorithms [9]:

- SOLT—Short open load thru, most used for calibration of coaxial systems. Calibration is not frequency dependent; it is quick and easy. The disadvantage is the reduced accuracy of measurement at very high frequencies.

Table 4.1 Types of network analyzer calibration [9]

Calibration type	Calibrated parameters	Use
Two-way two-port	S11, S12, S21, S22	Comprehensive calibration
Simplified two-pot	S11 and S21 or S22 andS12	Fast reduced transmission measurement accuracy
Single-pot	S11 a S22	Only reflection measurement
Frequency response	Selected S parameter	Parameter normalization

- SSLT—Short, short load thru the most common algorithm for calibration of the system for measurements on waveguides. The algorithm is frequency-limited, similar in property to SOLT.
- SSST—Short, short, short thru, an algorithm suitable for measurements of coaxial conduction and waveguides. It is designed for high frequencies. The properties are similar to those of SOLT, but the accuracy at high frequencies is increased. The algorithm is frequency-limited and requires very precise calibration fixtures.
- LRL—Line reflects line, an algorithm designed for high-quality coaxial elements or for surface waveguides. The advantage is to achieve the highest accuracy compared to previous algorithms. A high-quality transmission line is required and is limited in frequency.

4.4 Analysis of Antenna Radiation Characteristics

If measurements are required where it is necessary to irradiate the antenna with an electromagnetic wave, complications arise due to the reflection of the electromagnetic wave from the environment. This requires laboratory equipment consisting of an attenuation chamber, manipulators and instrumentation (Fig. 4.6).

These measurements are most commonly focused on radiation pattern measurements, polarization measurements, antenna gain, antenna directivity, phase, etc. The basic principle of these measurements is to measure the output voltage of the DUT. The device is irradiated with an electromagnetic wave from the transmitting antenna.

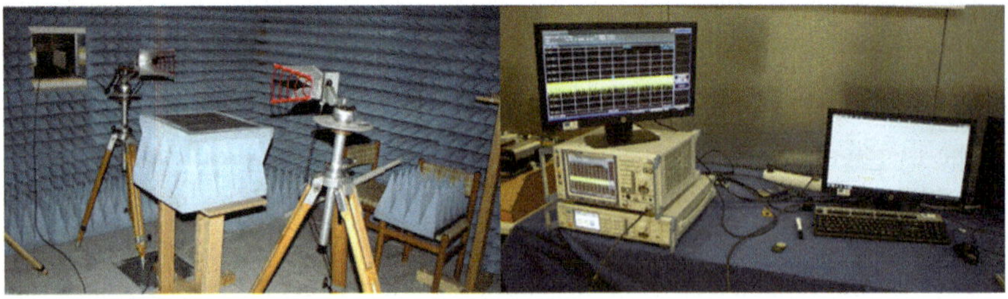

Fig. 4.6 Example of the use of elements in measurement

The parameters of the transmitting antenna must be known in order to determine the parameters of the DUT. The transmitting antenna can be used as a receiving antenna, in which case it receives the signal emitted by the DUT. This principle is mainly used for electromagnetic compatibility measurements.

Antenna measurements are made more difficult by the distance requirements between the antennas - as the size of the antenna increases, so does the distance between the antennas. The size of the measurement area itself may not be a problem, but it is necessary to ensure a non-reflecting environment, reduce noise to a minimum and eliminate external disturbance components of the electromagnetic field [10]. Particular attention must be paid to minimizing unwanted reflections from surrounding objects, the floor and the ceiling. This requirement is not met in the case of outdoor measurements. Open-area measurement conditions are suitable for deserts, dry plains or open spaces.

A more common environment in which real antenna measurements are made is the attenuation chamber [11]. The problem of antenna measurement can be summarized in several points: - unwanted reflections of waves from the environment must be avoided;

- take into account the necessary distances between antennas;
- measurement is more difficult if the antenna system is part of a
- larger objects (cars, planes, ships);
- the external environment may have an unstable and uncontrollable
- the surrounding environment and electromagnetic interference;
- indoor measuring workplaces cannot cover measurements of large
- antenna systems;
- it is necessary to provide costly instrumentation.

In the case of measurement of antenna radiation diagrams, the antenna under test is placed in an anechoic chamber and fixed on a computer-controlled manipulator. The measuring antenna is fixed at a certain distance and the position relative to the antenna is tested. The antenna to be tested is excited by the HF signal (39). At the measurement point (near or distant), the level and phase of the electromagnetic wave are captured by a measuring antenna.

Depending on which plane it is measured in, we distinguish [10]:

1. $E_\varphi(\theta = 90°, \varphi)$—$E_\varphi$ is a function of φ in plane XY
2. $E_\varphi(\theta, \varphi = 0°)$—$E_\varphi$ is the function θ in the plane XZ
3. E_θ $(\theta = 90°, \varphi)$—E_θ is a function of φ in plane XY
4. E_θ $(\theta, \varphi = 0°)$—E_θ is a function of φ in plane XZ

In ideal conditions for measuring the radiation diagram in a distant field and measuring gain, the antenna is irradiated by a uniform area wave. Wave has a flat wave front, where the vectors of the electromagnetic field propagate continuously throughout space [12]. For example, the vector E of a uniform area wave that propagates in the direction of + Z can be described using a one-dimensional wave equation [13].

$$E(z) = \hat{\rho}_w E_m e^{-ikz} \tag{4.7}$$

where $\hat{\rho}_w$ represents the polarization of the wave. In general, polarization depends on time and location, however, with a harmonic wave, the polarization vector has a complex value that must remain constant in the DUT region. The same applies to the magnitude E_m vector, which must remain constant throughout the DUT aperture. In practice, measurement in a distant field of radiation is used. The spherical wave generated by the source is approximated by a distant field and we consider it to be area-based [14].

4.5 Brief Analysis of the Near and Far Fields of the Antenna

For each antenna, it is possible to measure both the near and far field. The type of measurement is chosen according to the requirements. There are significant differences in price, site requirements, dimensions and measurement complexity between different measurements and methods. These parameters determine the suitability of a given measurement for a given situation and give preference to one type of measurement over another. Far-field measurements are more suitable for antennas operating at lower frequencies due to their simpler radiation pattern. Near-field measurements, on the other hand, are more advantageous for high frequencies because the radiation patterns of such antennas and arrays are much more complex and complicated, and it is possible to measure the polarization characteristics of the antenna [15, 16, 17].

Each type of measurement has an additional method of measurement which has certain advantages and disadvantages. This makes it difficult to compare techniques directly. A common advantage of all near-field measurements is that they can be carried out indoors. This eliminates problems with weather, electromagnetic interference from the surrounding terrain, security requirements and others. If we want to achieve a level playing field for far-field measurements, it is necessary to use a financially expensive anechoic chamber. Remote field measurements are often said to be less expensive. If we take into account the real requirements for outdoor remote field measurements, the financial costs increase significantly [18, 19].

After many years of development, near-field measurement has evolved to the point where it is considered to be the most important measurement of antenna characteristics [20]. The far-field measurement is not suitable for high-resolution measurements. Near-field measurement has been developed precisely to meet the needs of increasing accuracy, reducing price and the ability to diagnose the antenna in the shortest time with the greatest accuracy [21] (Fig. 4.7).

The radiation from the antenna consists of three regions. Transfers between these regions are not clear because the transitions between the regions are gradual [23]. The area closest to the antenna is the reactive near field. In the reactive region, the energy of the field decreases sharply with distance from the antenna. The average energy

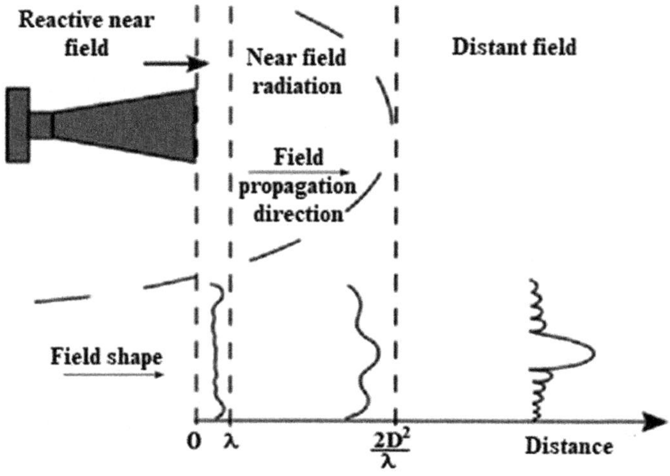

Fig. 4.7 Divisions of fields around the antenna [22]

density remains almost constant at different distances from the antenna, despite small fluctuations. When measuring the near field, the radiation in the reactive near field is measured and the far field is calculated using the Fourier transform. Near-field radiation propagates from the radiating area to a distance defined by 2D2/λ, where D is the largest aperture dimension and λ is the wavelength. Beyond this distance is the region of the far field, where the angular distribution of the energy does not change with distance, and the energy of the field decreases with the square of the distance [24].

The accuracy of the measurement is also affected by the dimensions of the room in which the measurement is made. The size of the test antenna and the dimensions of the room in which it is tested determine the critical angle φ. The calculated far-field radiation pattern will only be valid and accurate for the area ± φ. To determine the complete radiation pattern, it is necessary to measure the near field of the antenna at each angle, i.e., in spherical coordinates.

Three basic types of measurement are used. [22].

- Planar (planar)—the measurement plane is placed in front of the test antenna, which is stationary on the stand, and the near field is scanned by the near field probe, which moves in X and Y coordinates in front of the test antenna, forming a grid of the measured field.
- Cylindrical (cylindrical)—describes the cylindrical surface around the antenna. Unlike planar testing, the antenna is mounted on a rotating base and rotates around an axis. The near-field probe moves in an axis perpendicular to the axis of rotation. The rotation of the antenna and the near-field probe creates a cylindrical measurement grid.
- Spherical (spherical)—the spherical field around the antenna is measured over the surface of the sphere. The antenna to be measured is mounted on a rotating

stand that moves in two axes, and the near-field probe is fixed and directed at the intersection of the axis of rotation.

Antennas too large to be measured indoors can be measured outdoors. In this case, the test site must be adapted to the required measurement accuracy (Figs. 4.8, 4.9, 4.10).

Consistent preparation will ensure that the phase and amplitude distribution on the aperture of the antennas to be measured is uniform. Interference from a reflected signal with amplitudes less than 30 dB from the main lobe causes a gain error of

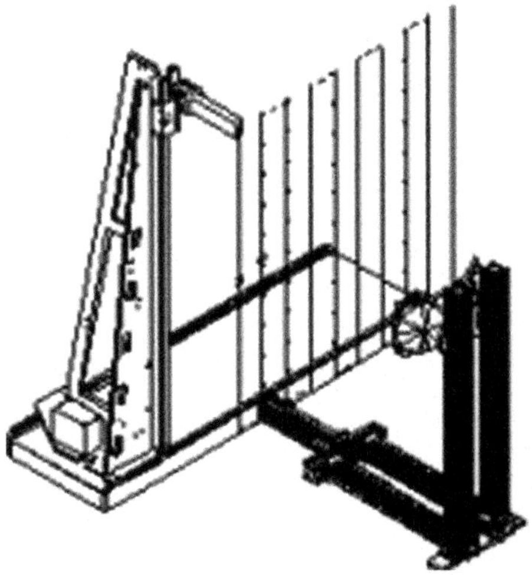

Fig. 4.8 Planar measurements [22]

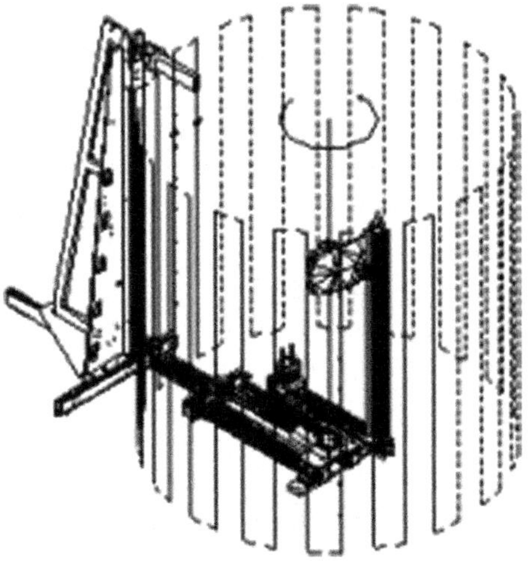

Fig. 4.9 Cylindrical measurements not [22]

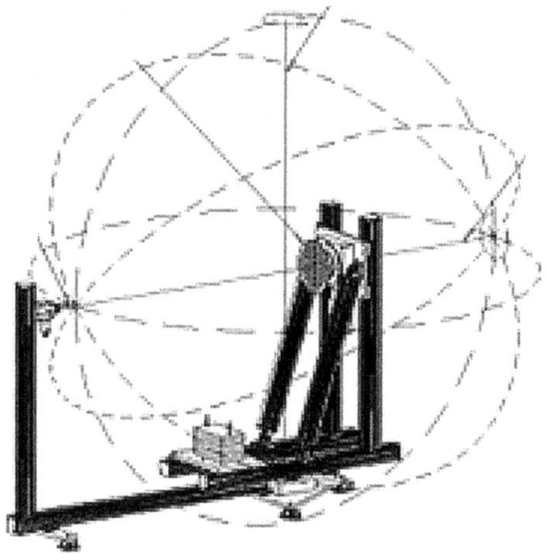

Fig. 4.10 Spherical measurement [22]

0.25 dB, which can cause significant side lobe distortion. Automating the measurement provides fast results, minimizes errors and gives good repeatability of the measurement.

For certain applications, it is more convenient or necessary to also measure the far field of the antenna to determine the amplitude or phase characteristics. For such measurements, it is advisable to use antennas with an operating frequency below 1 GHz. During the measurement, the transmitting and receiving antennas are placed at a sufficient distance from each other to simulate working conditions. The test antenna is irradiated by the measured antenna at a sufficient distance to produce an almost plane wave at the receiving aperture. The distance between the antennas may be defined as:

$$R > \frac{2D^2}{\lambda} \tag{4.8}$$

where R is the distance between the receiving and transmitting antennas, D is the size of the aperture to be tested, and λ is the wavelength [25] (Fig. 4.11).

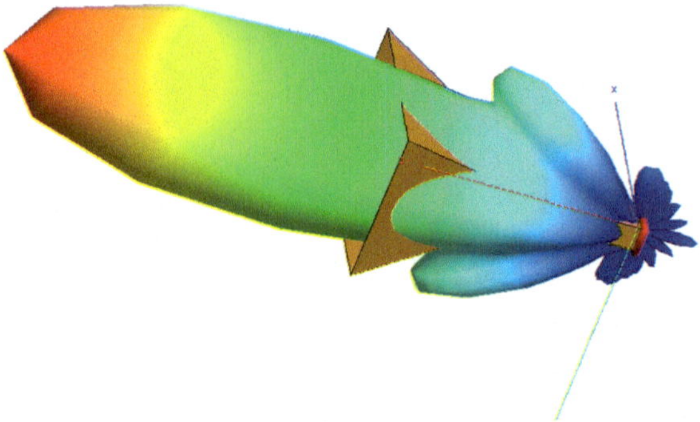

Fig. 4.11 Simulation of funnel antenna field—distant field

References

1. Kurokawa, K.: Power waves and the scattering matrix. IEEE Trans. Microw. Theory Tech.Microw. Theory Tech. **13**(2), 194–202 (1965). https://doi.org/10.1109/tmtt.1965.1125964
2. Rytting, D.: Network analyser error models and calibration methods. Dostupné na internete: http://www.uta.edu/faculty/jcchiao/class/EE3407/NA Error_Models_and_Cal_Methods.pdf
3. Davidson, D.: Computational electromagnetic for RF and microwave engine erring (2005). ISBN 9780521070126
4. Garg, R: Analytical and computational methods in electromagnetic (2008). ISBN 9781596933859
5. CHAPTER 3: ANTENNAS. Dostupné na internete: https://ocw.mit.edu/courses/electrical-engineering-and-computer-science/6-661-receivers-antennas-and-signals-spring-2003/readings/ch3new.pdf
6. Rodrigo, R.: Fundamental parameters of antennas. Dostupné na internete: https://www.ent.mrt.ac.lk/~ranga/courses/en4620_2010/L02.pdf
7. Besser,L., Gilmore, R: Practical RF circuit design for modern wireless systems, volume I: Passive Circuits and Systems, 2003. ISBN 1580535216.
8. Free space radio wave propagation. Dostupné na internete: https://www.iitg.ac.in/scifac/qip/public_html/cd_cell/chapters/a_mitra_mobile_communication/chapter4.pdf
9. Holloway, C.L., Mkenna, P., German, R.F.: EMA Inc. Lehman Chambers, on the application of computational electromagnetic techniques to the design of chambers for EMC compliance testing
10. Stutzman, Warren—THIELE, Gary: Antenna theory and design, 3rd edition (2013). ISBN 9780470576649
11. Introduction to network analyser measurements fundamentals background. Dostupné na internete: http://download.ni.com/evaluation/rf/Introduction_to_Network_Analyzer_Measurements.pdf
12. Milligan, T.: modern antenna design 2 John Wiley Sons Inc. Hoboken, New Jersey (2005)13 978-471-45776-3
13. Razavi, B.: RF Microelectronics, 2nd edn. (2012). ISBN 9780137134731
14. Navedtra: Antennas and wave propagation, naval education and training professional development and technology centre. ISBN 0504-LP-026-7580
15. Danek, K.: Moderní rádiový prijímač, kniha o jeho návrhu. Praha (2005)
16. Dobeš, J., Žalud, V.: Moderní rádiotechnika. Praha (2006)
17. Navedtra: Introduction to wave propagation, transmission lines, and antennas, naval education and training professional development and technology centre (1998). ISBN 0504-LP-026-8350
18. Joel, R.H.: Basic antennas—practical antennas and design. Amateur radio, USA (1994)
19. La, Y.T., Lee, W.: Antenna handbook fundamentals and techniques. USA (1993)

20. Antenna measurement theory 1. Introduction to Antenna Measurement. Dostupné na internete: https://www.keysight.com/upload/cmc_upload/All/ORFR-Theory.pdf?&cc=SK&lc=eng
21. Antenna fundamentals. Dostupné na internete: https://www.mtiwe.com/?CategoryID=353&ArticleID=163
22. Modelling and analysing polarization. Dostupné na internete: https://www.mathworks.com/help/phased/examples/modeling-and-analyzing-polarization.html
23. Antennas and propagation for wireless communication. Dostupné na internete: https://ecen665web.groups.et.byu.net/notes/ln2.pdf
24. Williams, A., Taylor, F.: Electronic filter design handbook, Fourth Edition. INTRODUCTION TO MODERN NETWORK THEORY, Chapter (McGraw-Hill Professional, 2006), Access Engineering
25. Černohorsky, D., Tichý, J.: Vyzařování a šíření rádiových vln a antény. Brno (1977)

Chapter 5
Technical Equipment of the Antenna Laboratory

The successful completion of the work requires sophisticated technical equipment. The measurements on which all the work is based are in the range of radio waves with frequencies below 10 GHz. Without the necessary equipment, it would be impossible to solve the work practically and to carry out the necessary experiments. The technical equipment is expensive and emphasis is placed on accuracy and quality of workmanship. The chapter contains a detailed description of the basic characteristics of the elements of the Aircraft Systems Laboratory. The description of the laboratory is based on a series of experiments designed for this purpose. Such a description will be the basis for future experiments, as it will provide information on the properties of the measured elements. The experiments were used to measure

- Characteristics of RF interconnect cabling with emphasis on attenuation,
- Characteristics of reference antennas,
- Characteristics of the attenuation chamber.

Part of the chapter is a brief description of the CADFEKO simulation environment used for the implemented simulations. For the need to obtain a large amount of measurement data, the control program for automating measurements in the LabView environment was analyzed and subsequently modified.

5.1 Antenna Laboratory Used in Experimentation

The laboratory is equipped for the measurement of antenna radiation characteristics. It contains a non-impact high-frequency chamber with manipulators, instrumentation, measuring reference antennas and measured antennas. In order to measure the influence of the fuselage on the radiation characteristics of antennas, scale models of aircraft have been produced with conductive surfaces on the surface of the aircraft model to enable these measurements.

M. Krchňák et al., *Advancements in Antenna Measurement*, Lecture Notes in Electrical Engineering 1108, https://doi.org/10.1007/978-3-031-48835-1_5

The radio frequency attenuation chamber built in the laboratory is a standard rectangular compact hybrid design. The manufacturer is Frankonia, model CHC. The chamber complies with the IEC/EN 61000-4-3 standard for electromagnetic compatibility measurements. The external dimensions of the chamber are 7.35 × 3.75 × 3.3 m. These dimensions allow measurements up to 3 m. The operating frequency range of the chamber is 30 MHz–18 GHz. The anechoic area is 1.5 × 1.5 m and starts at a distance of approximately 1.5 m from the rear wall of the chamber. The chamber is a hybrid structure, i.e. the walls of the chamber are made of 2 mm thick sheet metal with a wooden filling, sandwich type (Figs. 5.1 and 5.2).

A combination of ferrite plates and hybrid needles (absorbers) is used to absorb energy. The metal walls are covered with F006-type ferrite plates, which operate from 10 MHz to a frequency of approximately 1 GHz. For higher frequencies, hybrid absorbers of type H450s with a depth of 45 cm and a total area of approximately 37 m^2 are placed on the ferrite blocks. The maximum permanent electromagnetic field load is 1 kW/m^2 (600 V/m). The attenuation of the absorbers in the frequency range is 20 dB. The damping chamber contains its own electrical wiring, which is connected to the external distribution network via a series of network filters. The chamber is ventilated by a built-in air conditioner. The measurement can be monitored by a

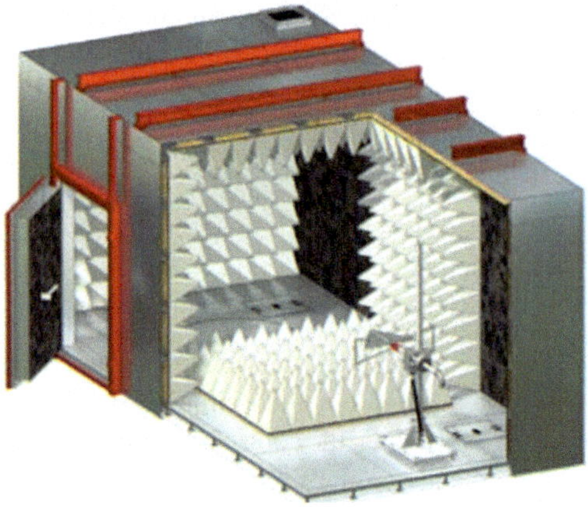

Fig. 5.1 Frankonia CHC compact anechoic chamber [1]

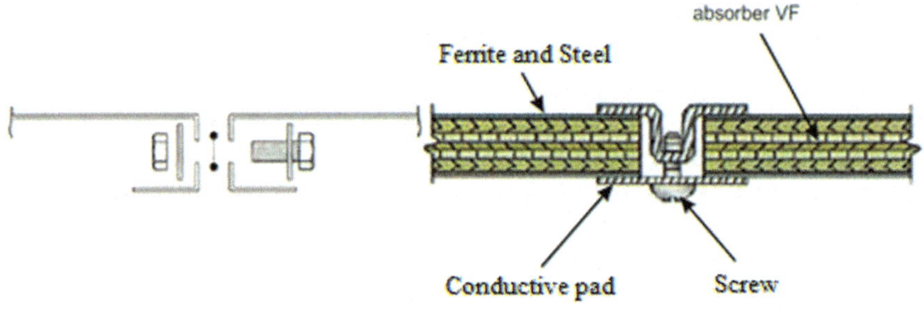

Fig. 5.2 Illustration of the construction of the chamber wall [2]

built-in battery-powered camera system. Connection of the internal equipment in the chamber to external equipment is via a panel with 6 × N connectors and a pair of optical cables (Table 5.1).

Two types of manipulators are placed in the chamber to manipulate the measured elements during the measurement. This makes it possible to measure the spatial radiation pattern of antennas.

The first type of manipulator is the Frankonia FAM4. It allows to move only in one plane—to change the height of the antenna in the range from 1 to 2.5 m and to change the polarization in two steps of 0 and 90°. The manipulator is connected to the control computer by means of an optical cable. The manipulator is mounted on a wheeled platform, which makes it easy to move around the chamber. The second type of manipulator is the custom-made ZTS vvu MSU01 manipulator. It is a stationary manipulator, i.e. it is impossible to change its position in the chamber. This manipulator is designed to manipulate the measured antenna. It has an extended movement compared to FAM4. Rotation is from 0 to 360° with an optional step. Elevation is from − 90 to 90°, also with an adjustable step. These movements are controlled by the LabView control software. The manipulator arm can be manually extended by about 50 cm from the retracted position.

The laboratory is equipped with instruments for RF measurements up to 13.6 GHz. The signal source for the measurements is the R&S SMA100 generator with a frequency range from 9 kHz to 6 GHz. The signal is measured using the R&S FSV 13 real-time spectrum analyzer with a frequency range from 10 Hz to 13.6 GHz. Two ETS Lindgren 3115 funnel antennas with frequencies from 750 MHz to 18 GHz are used as reference antennas. The maximum continuous load of the antennas is 500 W. Two instruments are used for measurements outside the chamber, the Anritsu s361e and the R&S FSH8, which can be used as a scalar network analyzer and spectrum analyzer up to 8 GHz. A computer with LabView control software is used to control and automate the measurements. It communicates optically with the instruments in the chamber via RS232 optical converters. Models of Boeing 737-600, Let L-410, MiG-29 and Mi-17 aircraft are scaled in the laboratory for measurements of antennas on the aircraft fuselage.

Table 5.1 Basic description of the anechoic chamber of the workplace [3]

Type	Fully anechoic construction
Absorbers	Hybrid absorber Franko sorb
Approximate dimensions	7 m × 3 m × 3 m
Power options	200 V/m < 1 GHz, 600 V/m > 1 GHz
Frequency range	30 MHz–18 GHz
Reflection	− 20 dB < 3 GHz; − 28 dB > 3 GHz

5.2 Laboratory Chamber Software Equipment

The laboratory is also equipped with the CAD FEKO simulation software. This is a set of tools designed primarily for the design and simulation of equipment (antennas, antenna elements, printed circuit boards, cabling, etc.) in the field of electromagnetic radiation. It is divided into two basic programme components. CADFEKO is used to design a model, define a material and create a grid for calculating and defining an electromagnetic field, and POSTFEKO is used to process the results of a simulation. The CADFEKO package includes a programme for designing the most common type of antenna, the ANTENA MAGUS. The programme also includes advanced methods for working with large objects, human tissue, various dielectrics and simulating electromagnetic compatibility. Based on the model, the program can calculate the behaviour of the model in an electromagnetic field. The results are interpreted in the form of 3D plots or 2D diagrams.

A network vector analyzer is best suited for measuring the characteristics of the test object over a wide range of frequencies. This instrument was not available at the workstation, so it was not possible to use it. One of the ways to perform such a measurement is to manually fine-tune the frequency of the generator and write the signal level from the instrument that measures the output level of the DUT. The LabView development environment speeds up the measurement. With its help, it is possible to compile a program for automating measurements. The program will allow control of the generator, spectrum analyser and manipulator, creating a program-controlled network analyser with automatic data acquisition. The programme will bring a number of advantages, e.g. it will increase the speed of the measurement itself incomparably, without the need for manual control of the instruments, and will allow automatic testing of the object in the specified frequency range.

The automation programme used is based on the "Automated Control Lvproj", project for automated control of stimulus profiles. Lvproj, which was created as a self-help within the framework of activities. The program allows to automatically measure the spatial radiation characteristics of antennas, i.e. the signal level depending on the angle of rotation and elevation of the antenna.

The program provides a good basis for various types of measurements in the attenuation chamber. The program "Automated Control Lvproj" is designed for measurements of a single frequency during the whole measurement. In the original version, it does not allow measurements in the specified frequency range. This program is primarily intended for controlling and collecting data from the R&S SMA100 and R&S FSVR13. Some of the proposed experiments required the use of the NI USB DAQ system because it was not possible to use the above instruments. The following modifications were made to the base programme: the possibility to choose a recording device (NI USB DAQ or R&S FSV13):

(a) in the case of DAQ, use the display of current–voltage on selected analogue inputs;
(b) in the case of using FSV13, there is a choice:

1. measurement at one specific frequency;
2. measurement of the frequency spectrum—spreading frequency within defined boundaries with the required step (software network analyzer);

(c) storing the measured data in a table in the form of a text file and editing the file header according to the type of measurement;
(d) modified display of the position of the manipulator in elevation;
(e) added options for choosing one of the present measurements, with the possibility of creating your own present;
(f) removed options for selecting half the antenna measurement space;
(g) increase the resolution of the frequency setting from 3 decimal places to 4 decimal places with a step of 0.0001 GHz (originally 0.001 GHz);
(h) measurement of the signal level by the spectrum analyzer occurs in ZERO SPAN mode, which leads to an increase in the measurement speed and an increase in the dynamic range. Initially, the program constantly searched for the maximum signal value using the MARKER SEARCH function in the monitored window in a given frequency range;
(i) reducing the measurement time by adjusting the waiting loop to 1/2 of the time without negatively affecting the measurement results;
(j) supplementing the measurement with the GABO measuring preparation for measuring the phase ratios of the signal on the antenna.

LabView stands for Laboratory Virtual Instrument Engineering Workbench. It is a development platform and environment based on the graphical programming language from National Instruments. The graphical programming language is called "G" (not G-code for cnc), originally from Apple's Macintosh. LabView is widely used for data acquisition, instrument control, and automation in industrial applications. The environment has broad support for operating systems including Windows, Unix, Linux, and OSX (Fig. 5.3).

The programming language works with blocks that resemble a flow chart (data flow type). The program is made up of a structure of graphical blocks that the programmer connects by means of conductive links and paths. They connect blocks and variables. The block function is executed when an incoming variable is at the input of the block. Although it appears that the program can execute many commands

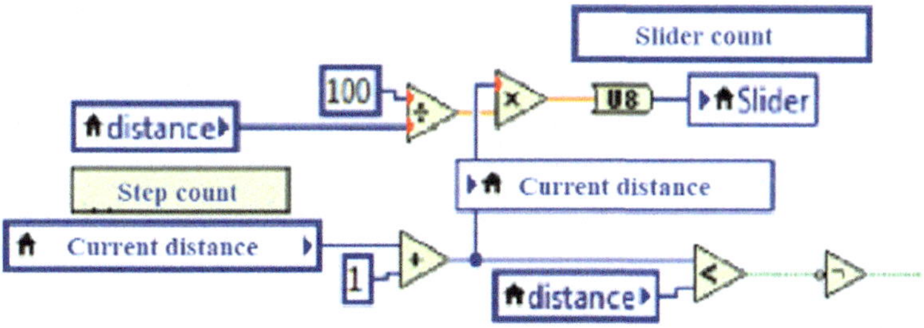

Fig. 5.3 Example of "G" graphical programming in LabView

at once, the code can process the data in parallel, block by block. Multi-processing and multi-threading are performed automatically by the built-in machine time organizer, which ensures the same program operation in different operating systems.

LabView allows you to create a graphical user experience for different applications or development cycles. The program consists of subroutines called "virtual instruments" (VI). Each VI has three parts: a block diagram, a graphical interface, and a VI connection panel. The last panel is used to link VIs together, creating a hierarchy of VIs as they are executed. The graphical interface is used for interaction between the user and the program, allowing the user to control the program.

The GUI contains various inputs in the form of switches, buttons, or rotary selectors, or outputs in the form of displays, fields, graphs, or bar graphs. Graphical inputs or outputs are connected to the program through blocks, in which their state changes depending on the action performed. The block diagram is used for the programming itself. Blocks are connected to each other in it, and it forms the program itself and determines its flow in time.

The biggest advantage of using LabView is practically unlimited support for devices such as gadgets, cameras and many others. Users can connect to a given hardware using pre-programmed environment blocks, or they can create their own blocks for communication with a given device, as the environment contains blocks for communication using GPIB, USB, RS232 and others. It also includes built-in blocks for communicating with hardware using MAX (Measurement and Automation Xplorer) and VISA (Virtual Instrument Software Architecture) tools.

The environment contains a code compiler that translates the block diagram into native code for the computer's CPU, which increases the performance and efficiency of the program. The result is a single file with compiled code. The code is executed using the LabView runtime engine, which contains some default code syntax and preprogrammed most frequently performed tasks. This reduces the compilation time of the program and ensures the universality of the code between the OS.

The LabView environment itself contains only a few basic math blocks. All other blocks necessary for programming must be supplemented with libraries. The latter is available in a huge range for various instruments, data collection and generation, Mathematics, Statistics, data processing, analysis, etc. Libraries are available on the official website of the manufacturer LabView, or on the website of the manufacturer of devices. Another advantage is the support for the MathScript programming language, which can be linked using scripting blocks with a graphical interface. MathScript is fully compatible with the MATLAB programming environment, which greatly expands the use of the program.

5.3 Laboratory Chamber Coaxial Line

High-frequency coaxial cabling forms the main way of connecting elements in the laboratory of aircraft antenna technology. With the help of coaxial cabling, the test and tested devices inside the chamber are connected to the measuring instruments that are located outside the chamber. This wiring feeds the signal to the transmitting antenna from the generator and the signal from the receiving antenna to the analyzer. It should therefore have the characteristics to transmit high-energy signals (of the Order of W) to the transmitting antenna and very low-energy signals (mW to μW) with as little attenuation as possible over the entire frequency range. From this, it follows that the characteristics of the wiring form an integral part of the measuring chain. Its properties negatively affect the measurement results and therefore it is necessary to know its parameters and include them in the results.

The Laboratory of Aircraft Antenna Technology uses LMR195 type cable with an impedance of 50 Ω and a total length of about 10 m. The dielectric is foamed polyethene, the inner conductor is a solid copper conductor, the dielectric is wrapped with conductive aluminium foil, and on it is an outer copper conductor in the form of a mesh. The maximum transmitted power specified by the manufacturer is 2.5 kW, which far exceeds the capabilities of the laboratory [4].

The integration of the workstation is shown in Fig. 5.4. The R&S FSH8 was used in network analyzer mode. The frequency range was set to our desired band of 800 MHz–8 GHz. The instrument was connected with two cables of 2 m length with N connectors to the connection panels of the chamber. The beginning of the wiring was connected to port P2 of the device, which in this case is the output of the generator. The end of the wiring was connected to port P1, which is the input of the analyzer. The parameter being measured is S21. Interconnection panels allow to connect devices inside the chamber with measuring instruments outside the chamber. The panel is another element in the signal path, but the influence of the connectors is very small, so it does not need special attention and its influence can be neglected. Inside the chamber, a 3 m cable connection was made using a connector. This created a direct connection between the output of the generator and the input of the analyzer. In the path of the signal is the cable + connector, i.e. the resulting waveform shows the signal loss on this chain.

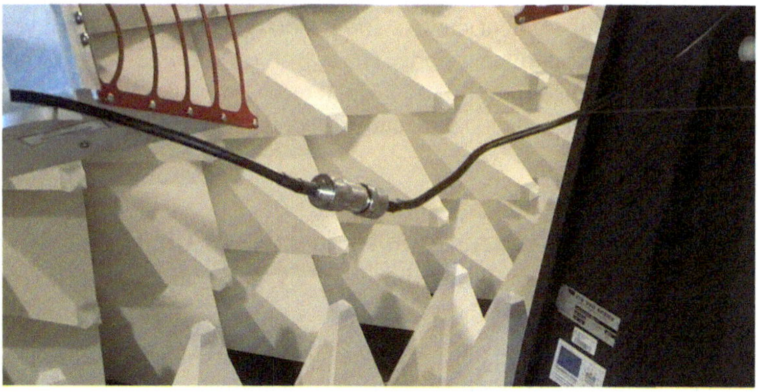

Fig. 5.4 Cable coupling connection details of cable coupling

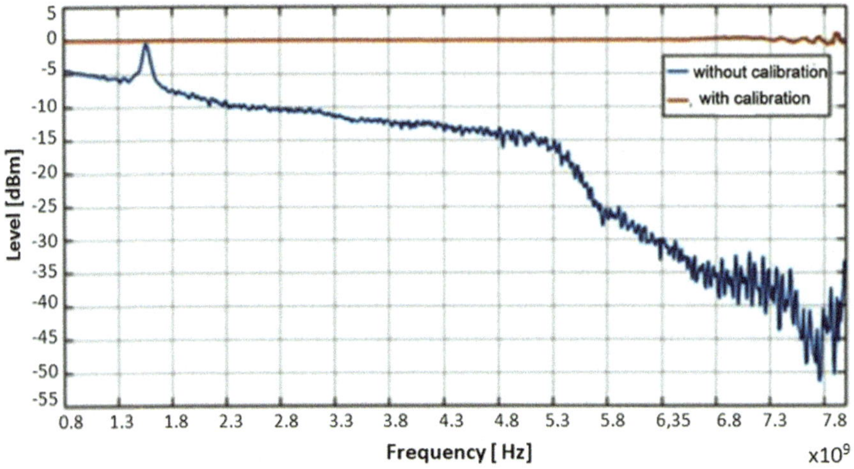

Fig. 5.5 Measured the attenuation of cabling in the laboratory

From the measured course (Fig. 5.5), a significant attenuation of the wiring used is noticeable. The cabling shows a signal attenuation of – 5 dB already at a frequency of 1 GHz. Cabling attenuation increases significantly. The limit frequency can be considered being the 5 GHz region, where a break occurs. Up to this frequency, the attenuation increases gradually, reaching levels of – 15 dB. At this frequency, it shows a sharp increase in attenuation. At frequencies from 7 GHz, the attenuation ranges in the – 45 dB region.

A calibration to this attenuation was then performed. The calibration principle is described in Sect. 3.3. All other measurements were performed with a valid calibration for the lead attenuation, i.e. the results are corrected for the given lead attenuation. This makes it much easier to measure and evaluate the results, as there is no need to deal with this problem further. The figure shows a significant effect of calibration. After calibration, the instrument shows a line at 0 dB, which represents a nearly ideal interconnection of the instrument ports. The waveform shows a slight ripple at the end of the frequency range from 7.3 to 8 GHz. It reaches a value of about 1 dB, which we can ignore since it does not significantly affect the result.

5.4 Reflective and Attenuation Properties of the Walls of the Laboratory Chamber

As described in the introductory chapters, the attenuation chamber creates a suitable measurement environment with defined parameters. In an ideal case, the walls should achieve infinite attenuation and a balanced frequency characteristic in the range of units of Hz to hundreds of GHz. However, this cannot be implemented. In a real

Fig. 5.6 Antenna distribution when measuring frequency response

environment, the walls of the chamber have a certain frequency response, i.e. is the working area of the chamber determined by the manufacturer. The zone of silence in the chamber is not perfect and there are reflections from the walls, but with a very small amplitude and therefore only minimally affect the results [5].

The experiment to verify the frequency dependence of the chamber walls consisted of measuring the level of the reflected signal from the chamber walls in a wide frequency band. ETS Lindgren antennas, model 3115, were placed one above the other on FEM4 manipulators at a distance of 1 m and a height of 1.4 m from the centre of the antenna. The distance was measured from the tops of the absorbers to the antenna aperture (Fig. 5.6).

The R&S FSH8 was used in network analyzer mode and set to measure the S21 parameter. The frequency range was 800 MHz–8 GHz. The instrument was calibrated for the effect of the coaxial line before the measurement.

The instrument used was set to average ten measurements. The resulting data were transferred to the PC and displayed using the MATLAB interface. Two sets of measurements were performed:

1. With the same polarization of the receiving and transmitting antennas, the antennas were vertically polarized.
2. With different polarization of the receiving and transmitting antennas, the receiving antenna was horizontally polarized and the transmitting antenna was vertically polarized.

Analysing the data from the measurement of the frequency response of the chamber walls (Fig. 5.7) and the measurements of the mutual transmission of the antennas at different antenna polarizations, two significant dips were detected. The first was at 4.38 GHz. At this frequency, the signal level dropped by 17 dB. The overshoot is most pronounced in this frequency range and is present in both the antenna measurement and the reflection measurement from the chamber walls. The second significant drop in the measured characteristic with different polarization of the antennas is at a frequency of 6.2 GHz. At this frequency, the signal level decreased by 23 dB. The second spike at the frequency of 6.2 GHz does not occur when measuring the characteristics of the antennas. When the characteristics are measured with the same polarization as the antennas, two significant dips occur again. However, these drops do not occur when measuring the mutual transmission of the antennas. The first significant drop occurs at a frequency of 3.98 GHz. The signal level dropped by 36 dB. The second overflow is at a frequency of 7.1 GHz with a signal level drop of 22 dB. The overflows detected have a very narrow bandwidth. The bandwidth of the overflow was determined when the level dropped by − 3 dB. The summary is shown in Table 5.2 (Fig. 5.8).

After conducting a more detailed analysis, we have assumed the presence of a common factor that influences both measurements and affects their results. A presumed cause of this phenomenon is the multiple reflections of the electromagnetic wave from the walls of the chamber to the measured area. The standard chamber layout has antennas placed perpendicular to the chamber walls. The electromagnetic wave can cause interference in the measured area by producing standing waves

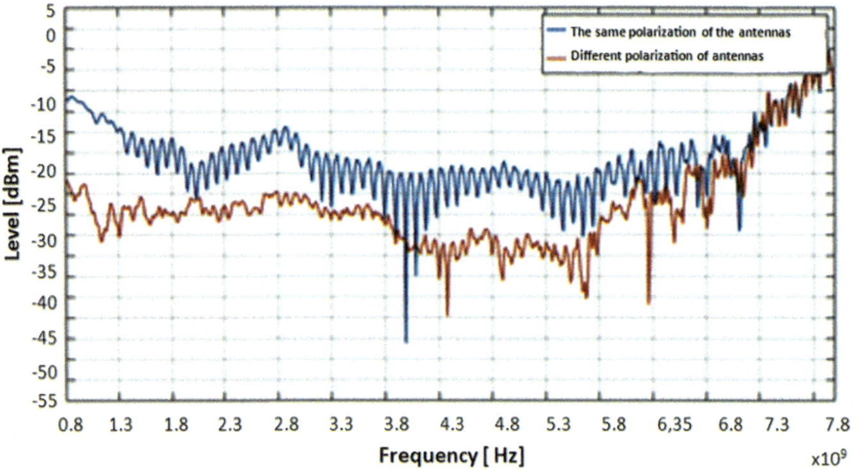

Fig. 5.7 Frequency characteristics of the walls of the anechoic chamber

Table 5.2 Drops in the level of the transmitted signal

Frequency (GHz)	Level drop (dB)	Bandwidth (GHz)
4.38	17	0.056
6.29	23	0.033
3.98	36	0.045
7.11	22	0.055

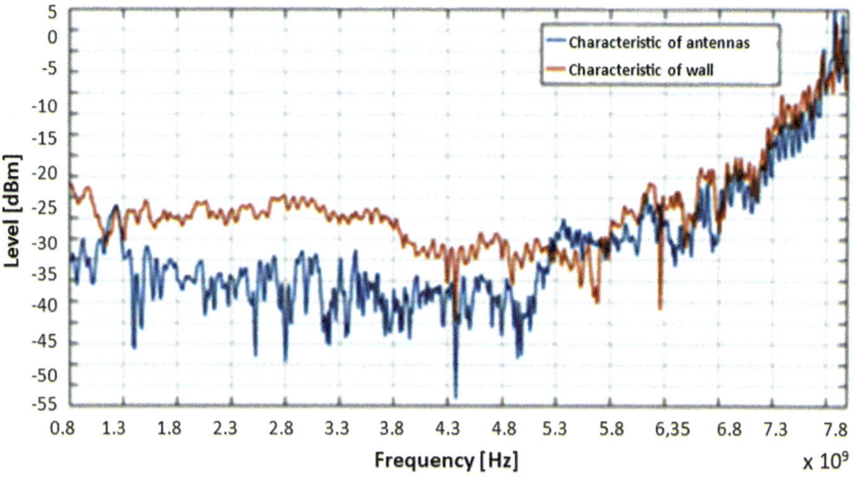

Fig. 5.8 Comparison of measured characteristics

through multiple reflections from the walls. This assumption can be easily verified or falsified through another experiment. In this experiment, the antennas were positioned at a 120° angle to the chamber wall (see Fig. 5.9).

The measurement principle was the same as in the previous instance. Plotting the measured data in a single graph provides information on how changing the antenna angle relative to the chamber wall affects the results.

Changing the angle of the antennas with respect to the wall prevented the expected phenomenon, which did occur in the case of the antennas having the same polarization. When using antennas with different polarizations, the opposite situation was observed. Two significant dips were observed when the antennas were oriented perpendicular to the walls, as described earlier in this section of the measurement. The expected phenomenon did not occur as in the case of the same polarization of antennas when the angle of the antennas was changed towards the wall. The

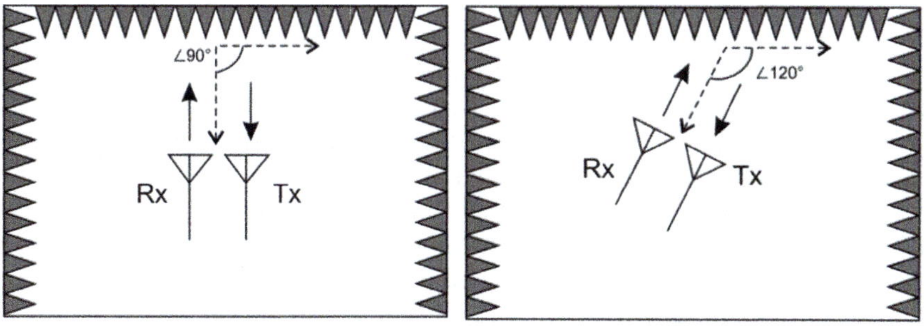

Fig. 5.9 Position of antennas measuring the frequency response of the wall

obtained characteristic does not filter out signal level drops, instead, it shows a fourfold increase in the number of drops and an increase in their decrease from 17 dB to about 32 dB. Thus, the chamber wall exhibits a sharp frequency dependence on the frequencies of the given overflows. The current sinks differ significantly from those identified in the mutual transmission measurements of antennas, allowing the effect of antennas to be disregarded. When conducting measurements in a chamber with antennas of different polarization, it is more convenient to arrange the measuring and measured elements so that they are perpendicular to the walls of the chamber.

By changing the angle of antennas on the wall, the expected phenomenon does not occur, as with the same polarization of the antennas. The opposite situation occurs in the measurement with different antenna polarizations. When the antennas were oriented perpendicular to the walls, there were two distinct overflows described at the beginning of this part of the measurement. By changing the angle of the antenna to the wall, the expected phenomenon did not occur, as in the case of the same polarization of the antennas. The characteristic curve obtained does not show a filtering of signal level overflows, but, on the contrary, shows an increase in the number of these overflows quadrupled and an increase of their decrease from 17 dB to about 32 dB. Thus, the chamber wall shows a very sharp frequency dependence on the frequency of overflow. Overflows are not identical or close to those identified in antenna overflow measurements, so we can neglect the influence of antennas. For measurements in the chamber with different antenna polarization, it is preferable to arrange the measured and measuring elements in the chamber so that they are perpendicular to the walls of the chamber (Fig. 5.10).

5.5 Transmission Properties of Antennas Measured in a Laboratory Chamber

The experiment aimed to determine the transmission frequency characteristics of the reference antennas within the wiring. Consequently, the antennas were used as references in the depolarization panel measurements. The measured characteristic allows for the quantification of the effect that the antennas have on the frequency characteristics of the depolarization panel. Funnel-type antennas, model 3115, manufactured by ETS Lindgren were used in this experiment. We designated them as references, based on their specified parameters. The catalogue data and characteristics provided by the manufacturer are obtained based on the specifications and procedures stated in the calibration standards outlining the methods for measuring electromagnetic interference between antennas. The designation of the standard is SAE ARP 958 and ANSI C63.5. The antenna manufacturer confirms the use of these standards in the antenna catalogue data. Both standards require Hertz dipoles or biconical antennas to be used [6] (Table 5.3).

Due to the unavailability of suitable laboratory conditions, measurement according to these standards is not possible, thus alternative methods have been

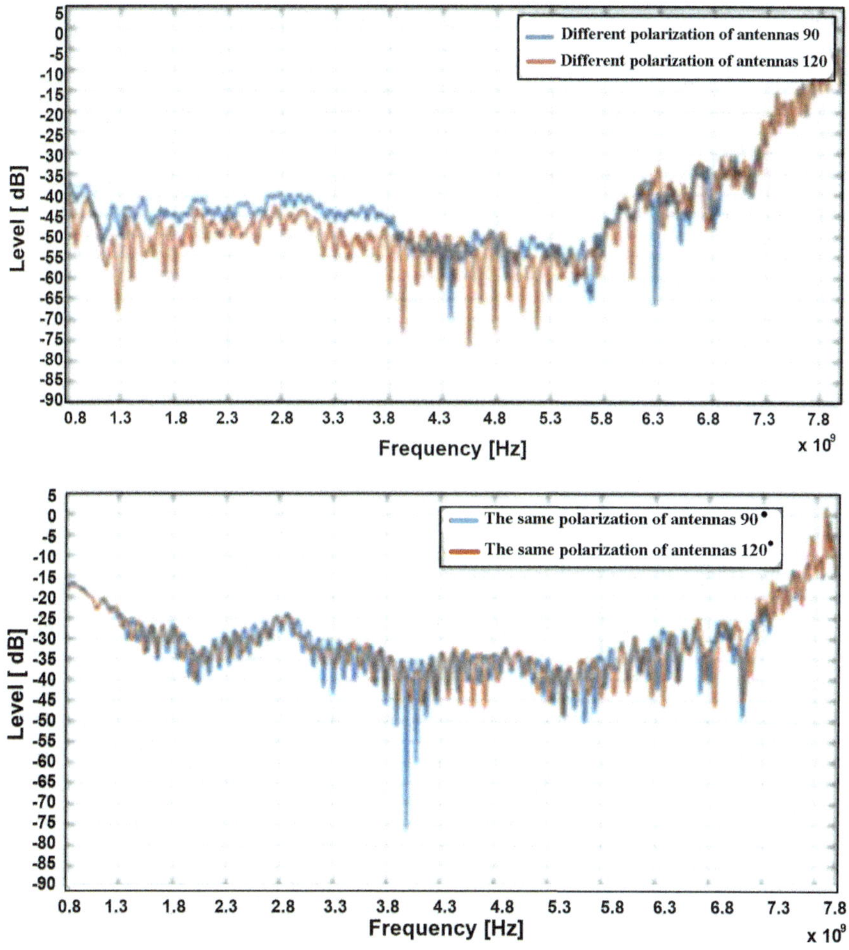

Fig. 5.10 Comparison of characteristics of the chamber wall

Table 5.3 Catalogue parameters of ETF antennas Lindgren model 3115 [7]

Catalogue parameters of antennas	
Minimum frequency	750 MHz
Maximum frequency	18 GHz
Connector	N-type
Nominal impedance	50 Ω
Maximum power	500 W: 800 MHz–1 GHz; 300 W: 1–18 GHz
Top performance	500 W
VSWR	< 1.5:1
Antenna type	Directional
Polarization	Linear

employed for evaluating transmission between antennas. The initial approach involves measuring the signal level using two identical antennas, following which the gain is recalculated following the formula denoted by the reference (5.1). The second method did not involve any conversion to gain. The waveform obtained illustrates the frequency characteristics within the range from 800 MHz to 8 GHz. This graphic representation portrays the individual signal level drops and peaks at varying frequencies, and this can be utilized to rectify the measured outcomes while evaluating the polarization panel.

$$G_T = \frac{1}{2}\left[20\log_{10}\frac{4\pi R}{\lambda} + 10\log_{10}\frac{P_r}{P_t}\right] \tag{5.1}$$

where

G_T is the gain of the antenna being tested (dBi);
R is the distance between antennas (m);
λ is the wavelength (m);
P_r power is received (W);
P_t is the transmitted power (W) [8].

We conducted the measurement using an FSH8 instrument that has a frequency range of 800 MHz–8 GHz and was calibrated for cable attenuation, as described in the previous section. Therefore, there is no need for any additional compensation in the calculation. We obtained input data by measuring two identical ETS Lindgren 3115 antennas facing each other. These antennas were mounted on two FAM4 manipulators, placed at a height of 150 cm from the floor of the attenuation chamber and 1 m apart. Before starting the measurement, we set the device to measure the parameter S12a and let it turn on for approximately 30 min. The results were obtained five minutes after the start of the measurement, to ensure that all values had stabilized. The measured data was also utilized for the second evaluation method. Catalogue profit values were derived by subtracting the estimated values from the corresponding graph [7]. Obtaining accurate data is not feasible, since it is not provided by the manufacturer.

According to Fig. 5.11, the profit measured by this method is similar to the catalogue data obtained from the available data. The antenna maintains an almost constant gain between the frequencies of 800 MHz–6 GHz. Because of the absence of catalogue data, the gain was only evaluated within the frequency range of 800 MHz–6 GHz. The verification of higher frequencies was not possible with catalogue data. The compliance of the measured gain with the catalogue data confirms the correctness of the method and the calibration of the instrument, which eliminates the influence of cable attenuation correctly, can be used for further measurements.

The second analysis involved plotting the antenna transmission data without any recalculation. The measurement and layout procedures were executed similarly to the previous case. Figure 5.12 displaying the waveforms, illustrates significant signal level drops that occurred at a different polarization of the antennas. They are visually emphasized on the chart by a green ring. Measurement overflows do not arise

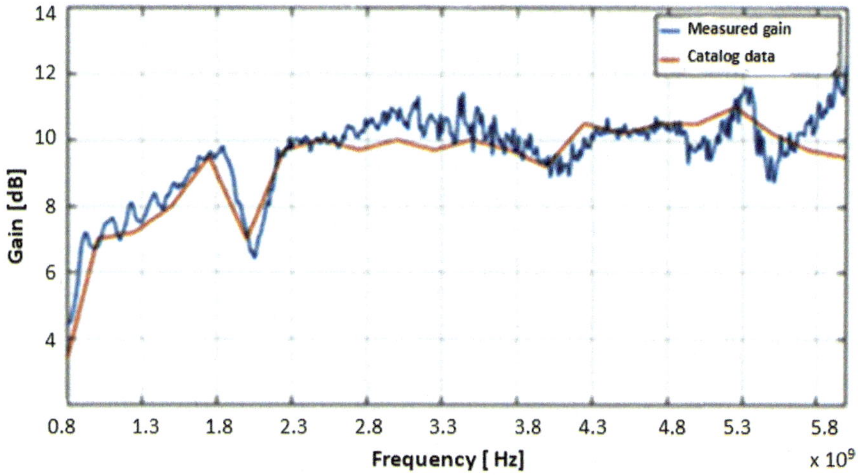

Fig. 5.11 Calculated and catalogued antenna gain ETS Lindgren model 3115

while measuring the signal reflection from the chamber's walls. A comprehensive analysis of the measurements presumes that the dips result from various factors that impact the measurements. These factors include multiple wave propagation in the chamber, multiple signal reflections from the walls, and the antenna's design. Further measurements will not involve addressing these dips. Instead, they will only be used for result compensation and avoiding misinterpretation of data.

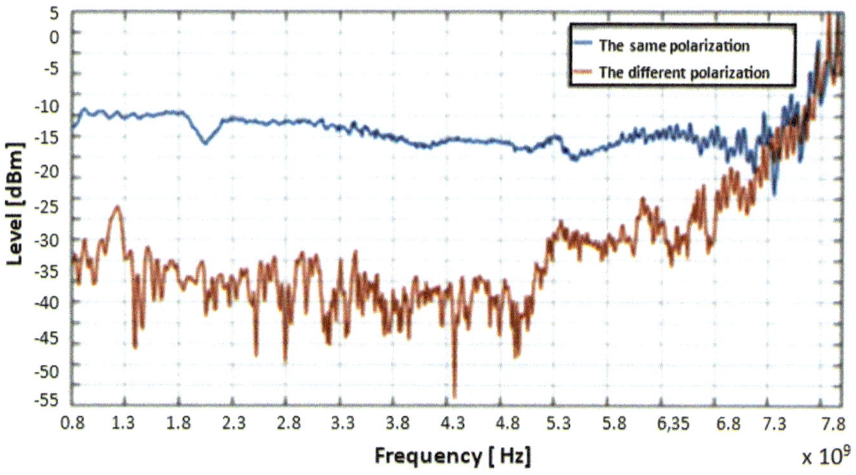

Fig. 5.12 Measurement of mutual transmission of antennas

5.6 Nearby Antenna Field in the Radial Direction Measured in the Chamber

To measure near-field radiation, antennas must be slightly slid or rotated during measurements. Manual measurements are challenging due to the difficulty in maintaining a consistent step throughout the measurement path. This is why a measuring system was designed to measure near-field radiation. The system consists of hardware components, namely: a motorized truck to shift the antenna, a converter for motor control, an RF measuring instrument, and a signal generator for the antenna.

The initial measurements for the depolarization panels were focused on designing a system to measure the field path from the antenna in the attenuation chamber. Realistically establishing the course of the radiated field in the attenuation chamber with increasing distance between the transmitting and receiving antenna was necessary.

Although not mandatory, it is advisable to know of the course, specifically how the signal level changes when the distance between the antenna's changes. In certain scenarios, the findings can specify the maximum distance between the antennas where it remains feasible to stimulate the receiving antenna using a signal from the transmitting antenna. The measurement had to determine the irregularities in the path, which could arise from the reflection from the chamber walls or the manipulator's influence. Mathematically determining this level drop makes identifying deviations effortless.

5.7 Hardware Equipment

For the initial measurement, a 20×40 mm wooden prism was used. Wood has less impact than commonly available plastic types on the measured characteristics. Although PTFE would be the best material for this use, the production of a profile with these dimensions would be very costly. A bracket was created to fix the antenna to this wooden prism (Fig. 5.13).

The antenna was manually adjusted using the scale attached to the prism. After each adjustment, it was necessary to record the field's magnitude manually from the spectral analyzer. This procedure was significantly time-consuming. Manually moving the antenna caused non-linear movement distance and rotation in other axes during each adjustment. As a result of this procedure, the data measured had an error, and the resulting graphs were randomly affected by this error.

To minimize errors caused by manipulation, we constructed a motorized vertical carriage to hold the antenna. In the LabView environment, we use the proposed Control Program to set the required measurement step and maximum distance from the transmitting antenna before initiating the measurement. Motor movement guarantees that each step is consistent and the data collection system ensures that the sample

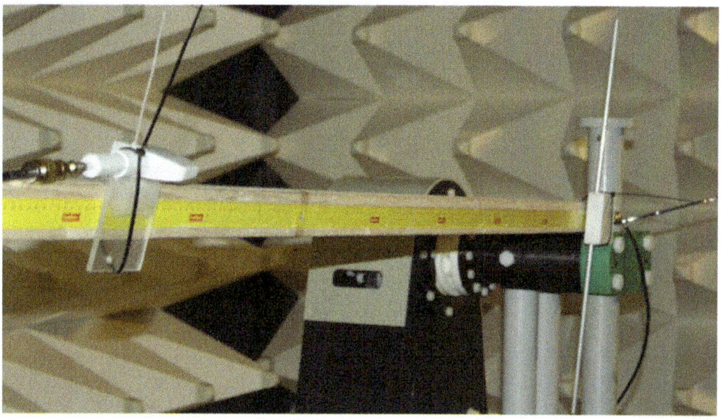

Fig. 5.13 Initial implementations of measuring workplace

is collected only after completion of the movement. The DAQ system provides manipulator control. The NI USB6210 is a multifunctional data acquisition device with a high sampling rate.

It enables measuring voltages at inputs, generating DC voltages with high resolution, generating PWM signals and slow AC waveforms, directly sensing voltages on thermocouples, and measuring resistance, current, frequency, noise level, acceleration, force, pressure and torque via USB (Fig. 5.14).

Provides:

1. 16 analogue inputs, resolution 16 bits, sampling rate 250 pcs/s;
2. 4 × digital inputs;

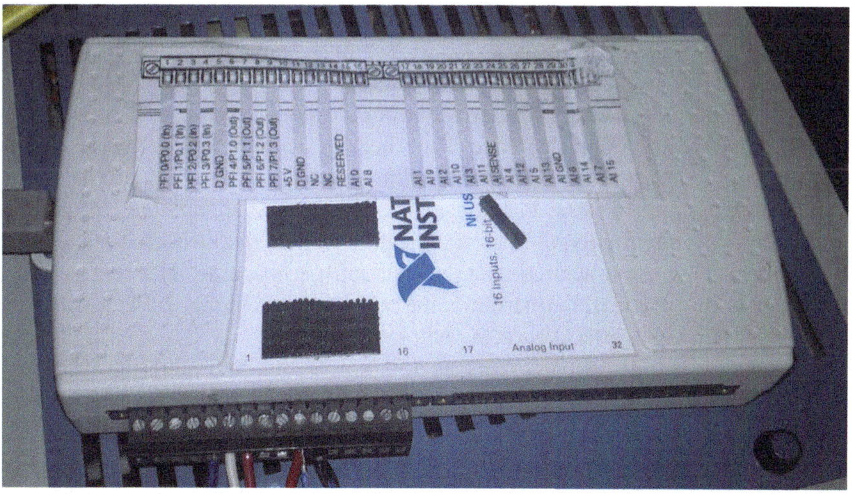

Fig. 5.14 NI DAQ converter

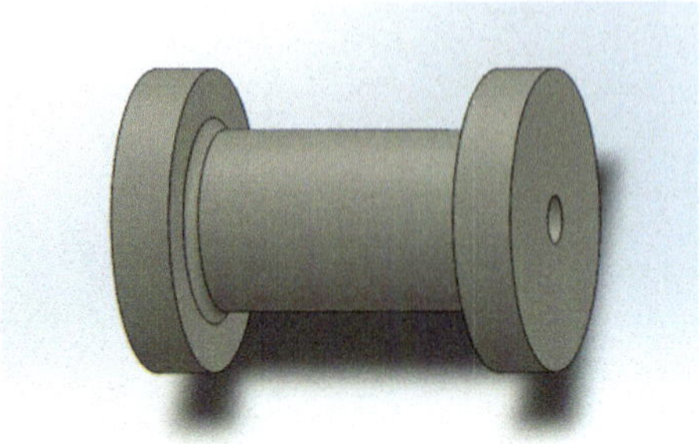

Fig. 5.15 Silo support wheel 3D model

3. 4 × digital outputs;
4. 2 × 32 bit timers;
5. power supply via USB;
6. full compatibility with LabView NI-DAQmx.

The converter was used due to the ease of operation via LabView. 4 digital outputs and 2 digital inputs are used. With the help of digital output, a signal is generated by the control board of the stepper motor.

The vertical manipulator consists of a power unit—a stepper motor and a gear rubber belt, control electronics for the stepper motor, a trolley for fixing the antenna, and wooden wiring.

The trolley carrying the antenna to be measured consists of three basic parts:

1. Support wheels;
2. Front;
3. Back.

Since these are measurements of electromagnetic waves, the trolley cannot contain metal parts. This fact was verified by using metal screws as wheel axes. When measuring, it was found that these screws negatively affect the measured results. Metal screws were replaced with plastic ones (Fig. 5.15).

The support wheels were rotated away from the silo to ensure they rested on the wooden prism with enough clearance. The wheel's axle is created by a screw linking the front and rear of the cart. The front and rear sections of the cart are built from 5 mm thick plexiglass parts. Plexiglass is readily available, economical, perform well, and most importantly, did not significantly impact the measured results. The components were manufactured using a CNC milling machine. The front and rear sections are not identical. Furthermore, the rear section is equipped with two holes in the axle, utilized to affix the towing belt to the engine.

Both the front and rear parts measure 150 × 150 mm. Self-adhesive Velcro is a reliable solution for attaching antennas. Using self-adhesive Velcro enables simple adjustment of the antenna position and provides reliable mechanical strength. The cart moves thanks to the 'timing belt', illustrated in Fig. 5.16. The belt is made of neoprene rubber and is reinforced with Kevlar in the direction of tension. The inner side of the belt features teeth at 2 mm intervals.

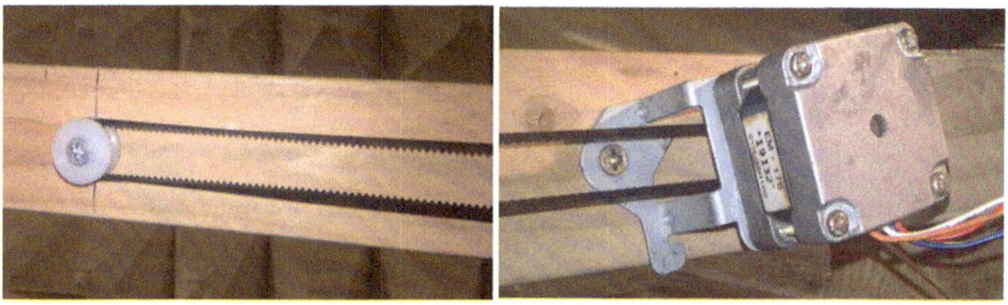

Fig. 5.16 Gear belt for deriving a trolley

The belt measures 6 mm in width and 6 m in length. Its limited flexibility renders it suitable for use at such lengths without compromising accuracy. The shaft of the motor is equipped with a gear that perfectly interlocks with the belt's one gear. This arrangement enables precise movement of the carriage even with small motor steps. The belt's use eliminates the possibility of a step being skipped (i.e. the belt slipping on the motor shaft); hence, the motor shaft rotates without moving the carriage.

The cart's drive unit constitutes a unipolar stepper motor. The toothed guide moves the trolley linearly along the rail. To control its movement, a Pololu A4988 integrated circuit is used for the motor. The motor is labelled EPSON 2007492 EM-178 STH-39H105. This is a two-phase bipolar motor with the centre of the windings drawn out, operating at a supply voltage of 12 V and with a maximum current of 1 A through the windings. The motor comprises 200 steps per revolution, equivalent to 1.8°/step (Fig. 5.17).

There are two basic wiring windings for stepper motors—bipolar and unipolar. The unipolar motor has two windings per phase, one for each polarity of the magnetic field. In this arrangement, the magnetic field can be reversed without changing the current flow through the winding, making the control circuit simpler. Each winding requires only one transistor. The most common winding arrangement can be seen in Fig. 5.18. One winding end is common, resulting in 3 wires for the phase and 6

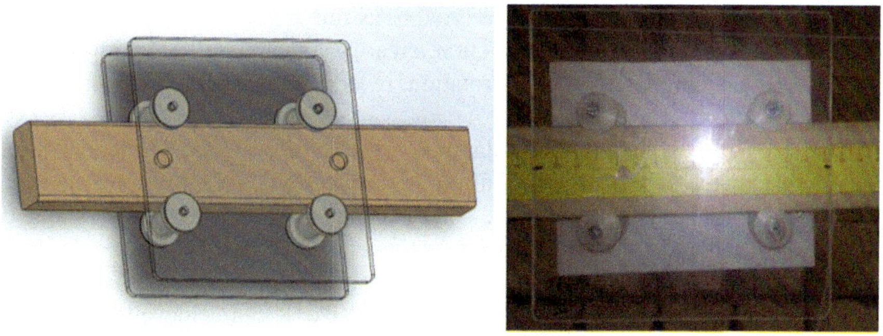

(a) **(b)**

Fig. 5.17 Antenna trolley **a** 3D model, **b** photo of the assembled trolley

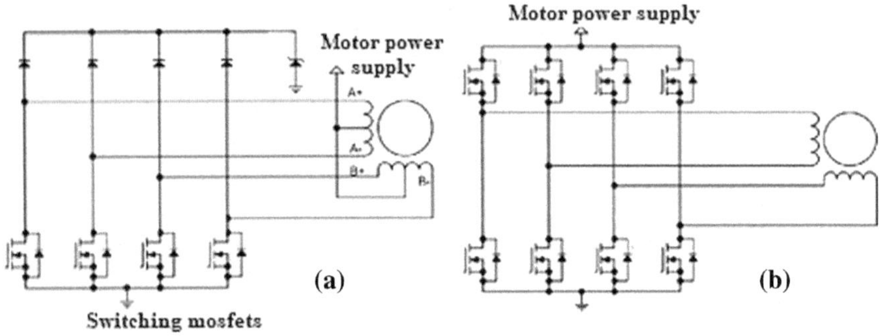

Fig. 5.18 **a** Unipolar engine, **b** bipolar engine

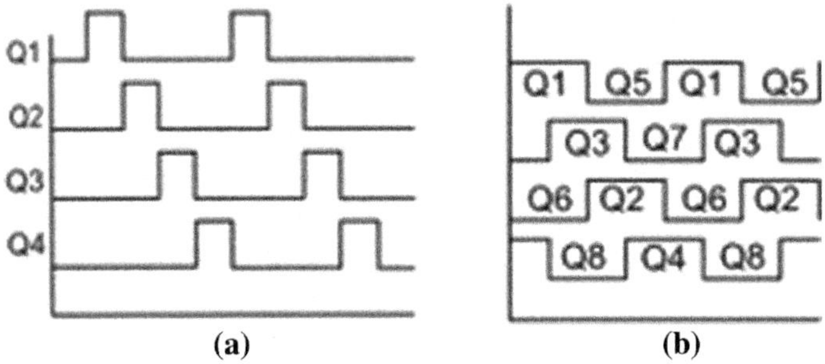

Fig. 5.19 The order of fastening of the motor windings **a** unipolar, **b** bipolar

for the entire engine. The windings' common connection is frequently established directly in the engine design, resulting in only one common conductor for the winding centres. Figure 5.19 shows the control circuit switching individual transistors in a specific order.

For a unipolar motor, gradual switching of the windings is used. The windings are disconnected and connected sequentially, starting with the first. One step is performed by the rotor for each pulse. This is referred to as full-step mode. The current flows through the winding in a single direction, creating a magnetic field with only one polarity. To change the direction of rotation, reverse the sequence of clamping the phases.

The second type of motor is a bipolar motor (refer to Fig. 5.18b). The latter has one winding dedicated to each phase. Controlling the electronics is more complicated as it requires reversing the current in the coil to change the orientation of the magnetic field. Usually, the H bridge is used for controlling the motor. The motor has two independent output wires, and the windings are not interconnected. Because the motor has better-arranged windings of the same size, it can achieve higher power output than a unipolar motor. Controlling bipolar motors requires a more complex logic than unipolar ones. Refer to Fig. 5.19b for the sequence of winding clamping. Specialized circuits, such as Pololu A4988, L293D, MC3479, SN754410, and L9942 are used to control bipolar motors. To switch and time, an array of MOSFETs in an integrated case and a microcontroller or control pulse generator are typically required. The unipolar motor can be connected to bipolar mode, but the reverse is not possible. The adjustment involves disconnecting the common centre of the windings to achieve one winding per phase. This knowledge was applied in the EPSON engine used.

To control the motor, we chose a control board with an A4988 circuit. The circuit allows for complete control of a bipolar stepper motor. In addition, it includes essential protection features like overcurrent and short-circuits protection, thermal fuse, and continuously adjustable output current. It also allows the motor to be controlled in five different modes: full step, half step, quarter step, and eighth step. The motor is being controlled in full-step mode, as a smaller motor step is not required. The supply voltage ranges from 8 to 35 V, and there is a maximum output current of 1 A without the need for cooling. The circuit has a significant advantage in the form of an integrated signal converter. The circuit controls the switching of motor windings based on the number of input pulses received. Each input pulse signifies one step. The impulses are transmitted sequentially over a single wire. By following this approach, the need for phase sequence tables or complex control programs is eliminated. A pair of H bridges with N-channel DMOS FET transistors forms the power output for the windings. The output current of the bridges is regulated using PWM with a constant OFF time. The shunt resistors and the reference voltage Vref determine the current.

Upon circuit startup, the DAC and all signal phases are set to the initial position. When a pulse is detected in the STEP input, the signal translator adjusts the output level of the Digital-to-Analog Converter (DAC) and modifies the phases of the H-bridge control signals in accordance with the integrated sequence table to initiate motor movement. The DIR input is utilized to adjust the direction of the motor rotation. This change in the rotation direction occurs exclusively at the next rising edge of the Step input signal (Fig. 5.20).

This circuit is designed to control a bipolar motor, but the EPSON motor is of the unipolar type. There are two possible solutions to this problem: The first solution is to

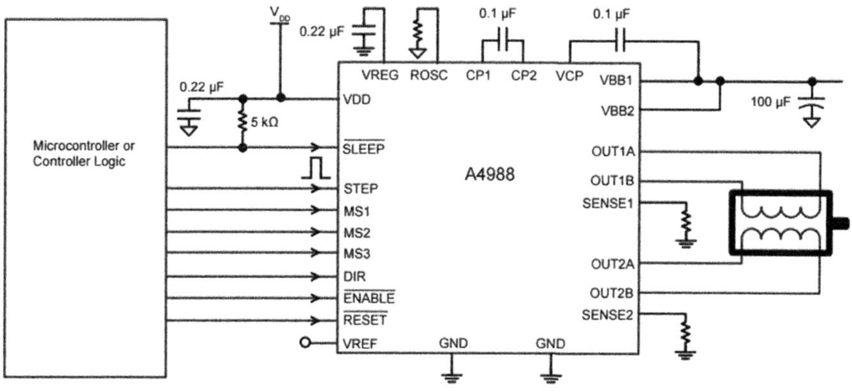

Fig. 5.20 Recommended circuit wiring [9]

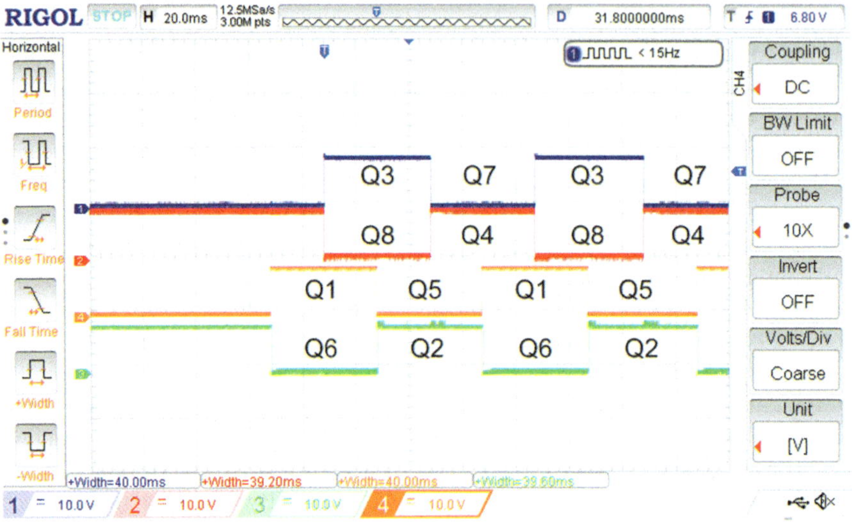

Fig. 5.21 Control circuit input signal

disassemble the motor and disconnect the common centres of the windings, creating a quasi-bipolar motor. This action results in a quasi-bipolar motor. The second option is easier.

Connect the circuit outputs to the 4-Channel Oscilloscope to observe the individual output activation order. The measured order can be marked as in Fig. 5.21 based on the theoretical course of Fig. 5.19b. Therefore, the phase order and transistor switching are maintained. Since the order of the circuit's outputs does not match that of the oscilloscope's inputs, Qx's order is displayed differently than in Figs. 5.19b and 5.22.

After examining the course more closely, it becomes apparent that it shares similarities with Fig. 5.19a, specifically the sequence of phases necessary for controlling a unipolar motor. Moving individual waveforms on the oscilloscope screen will produce the desired sequence of signals for controlling the motor in use.

Measurements indicate that the control IO should also be capable of controlling the used EPSON motor. The obtained results can be further validated by inspecting the internal circuitry, particularly the MOSFET H bridge, to observe which transistor is open or closed. Upon connecting the motor to the control IO, it began to rotate according to the input pulses and set rotation direction. This simple modification made it possible to use the A4988 circuit, even though it was not designed for the specific type of engine in use.

Following the confirmation of control IO and motor functionality, the base board with connectors for the motor, limit switches, and power supply was designed. The board features an RJ45-type connector for the input signal. The DPS is connected to the output through terminal blocks (Fig. 5.23; Table 5.4).

Since the motor has a non-negligible inductance, its electromagnetic field will negatively affect the measurement chain, causing a measurement error. For this

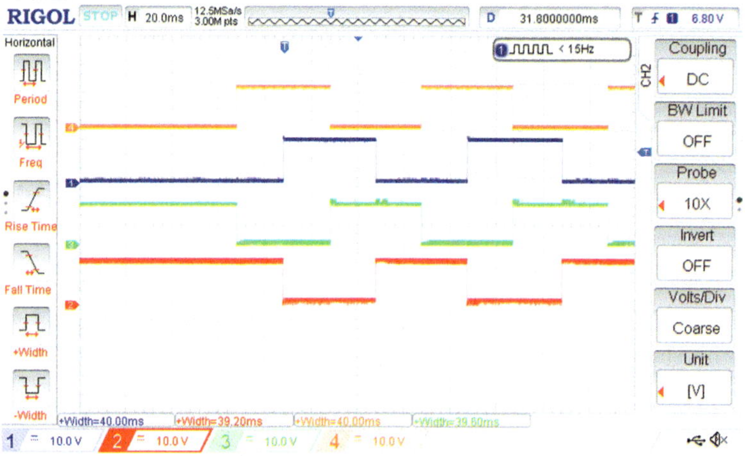

Fig. 5.22 Changes the order of the outputs of the control IO

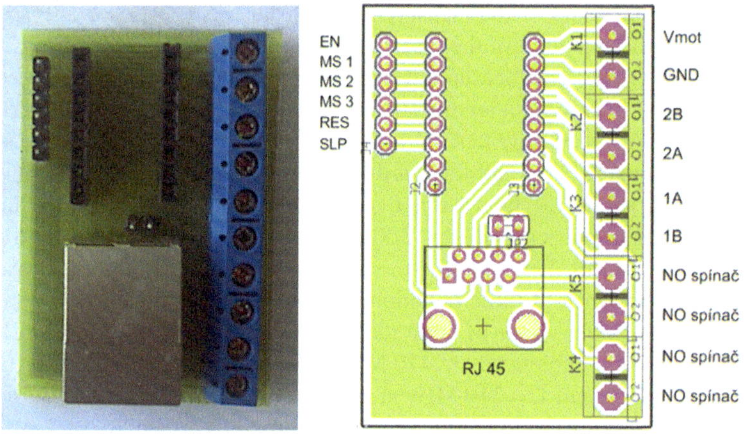

Fig. 5.23 Steering motherboard and motor connection

Table 5.4 Motor connection

Driver	Outlet on the board
Orange	2A
Red	1A
White	1B
Blue	2B

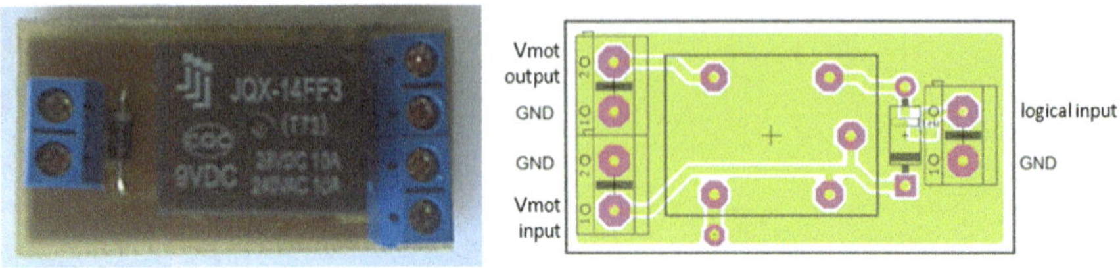

Fig. 5.24 Spine relay for motor supply voltage

reason, the supply voltage to the motor is repeatedly switched off during the measurement. This shutdown is performed by a simple relay controlled by the DAQ system (Fig. 5.24).

The Hardware is only one of the most important parts of the measurement chain. No less important is the operating software. It handles the entire management of the measurement workstation. It communicates with the hardware via a National Instruments DAQ converter, which then sends signals to the hardware. It also takes care of the initial setup of the measurement instruments (generator and spectrum analyzer) and the sampling of the measurement signal from the instruments via the GPIB/USB interface. The measured data are stored in tables for further processing. All functions of the program are interwoven through the graphical user interface.

The control software is written in LabView. This program provides direct support for instruments using GPIB and also has direct National Instruments DAQ converters.

Figure 5.25 shows the field name displayed by this system. The measurement is performed at frequencies of 210, 300 and 410 MHz. At these frequencies, the wavelength is 1.42 m, 1 m and 75 cm, so it can be measured in a wide range and move within the range of three radiation regions, see Fig. 4.7. The measurements have been made in the range of close radiation, to a distance of less than 2λ, with the limit being a high frequency of 410 MHz, where the limit is 1.5 m. The far field is located in the case of $f = 210$ MHz at a distance of about 3 m from the radiation source, which is the maximum length of the measurement chain.

At 2.4 GHz, the measurement range in the near field would be limited to a distance of 24 cm, so it was necessary to measure with a very small step to obtain the same number of measurements as at the frequencies used.

According to theoretical assumptions, the field level should decrease by the square of the distance from the signal source. It can be seen that when the frequency is increased to 410 MHz, the signal level is almost constant over the entire measured distance, varying only in the range of 0.0002–0.001 V. The significant effect of distance on the signal level is seen mainly at the low frequencies of 300 and 210 MHz. Here, a gradual transition from the reactive zone to the Fresnel zone of radiation can be observed, and the magnitude of the field varies over a wide range. This transition is not pronounced at 410 MHz, which means that at 2.4 GHz it would not be measurable with the given laboratory preparation. This measurement provided information about the field level as the distance between the antennas increases. From the measured data, it can be concluded that at a frequency of 2.4 GHz, it is small along the entire

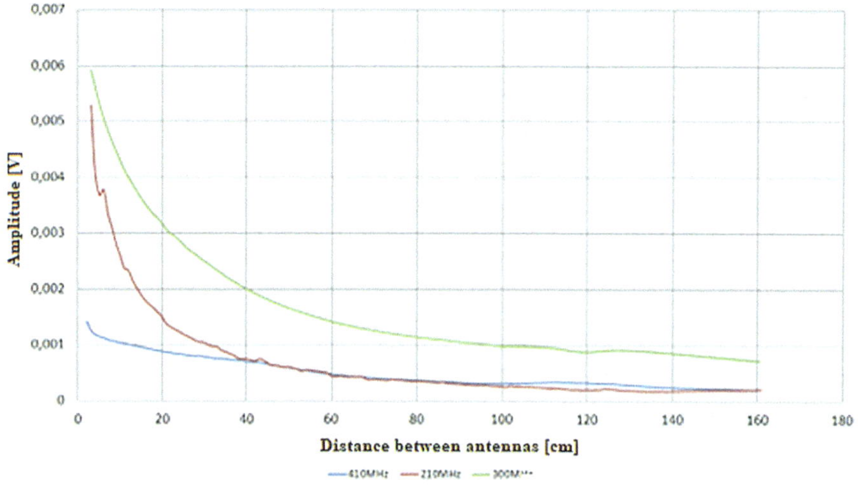

Fig. 5.25 The course of the electromagnetic field in the anechoic chamber

length of the signal, almost constant in magnitude, and varies only in a small range. The transition between the reactive zone and the Fresnel zone does not affect the results, as it is not measurable in the given case.

References

1. Antenna feed impedance. Dostupné na internete: https://www.radioelectronics.com/info/antennas/basics/antfeedimpedance.php
2. Anechoic chamber. Dostupné na internete: https://i.pinimg.com/736x/1e/22/bc/1e22bc6511468a66e0926e885b93aae2--anechoic-chamber-science-projects.jpg
3. Žalud, V.: Vysokofrekvenční prijímací technika. Bratislava (1986)
4. DOUBLE-RIDGED GUIDE 3115 Double Ridged Guide Antenna. Dostupné na internete: http://www.ets-lindgren.com/dataplate/antennas/double-ridged-guide/4002/400203
5. Képeši, V., Labun, J.: Radar signal attenuation due to finite radome thickness. Naše More **62**(3) (2015). ISSN 0469-6255
6. Labun, J., Soták, M., Kurdel, P.: Innovative technique of using the radar altimeter for prediction of terrain collision threats. J. Am. Helicopter Soc. **57**(4), 045002-1–045002-3 (2012). ISSN 0002-8711
7. Matzner, H., Levy, S.: Antenna Gain. Dostupné na internete: http://www.hit.ac.il/.upload/engineering/antenna_-_exp4gain.pdf
8. User's Manual Suite 7.0, EM Software & Systems-S.A. (Pty) Ltd, 2014. Dostupné na internete: http://altairuniversity.com/wp-content/uploads/2015/03/UserManual.pdf
9. LMR®-195 Flexible Low Loss Communications Coax. Dostupné na internete: https://www.timesmicrowave.com/documents/resources/LMR-195.pdf

Chapter 6
The Optimalization of the Depolarization Panel

In the second chapter of the monograph, the theoretical principle of polarization and reflection of an electromagnetic wave on a wire grid and a sheet is explained in detail. Based on the theoretical assumptions, an experimental depolarization panel has been proposed in the development process. The panel uses the reflection and decomposition of waves from conductors and conductive surfaces to achieve a depolarization effect [1–3]. The experimental panel was designed with the requirements of flexibility, i.e. the ability to change its parameters. Basic measurements were made on the experimental panel to provide basic information for the design of the panel with the required characteristics. In order to quantify these effects, a series of experiments were carried out:

- Influence of conductor spacing on wave reflection,
- Influence of conductor diameter on wave reflection,
- Effect of conductor orientation on wave reflection,
- The effect of wire settling on the resonant frequency of the panel,
- The effect of changing the distance of the plate from the wires on the resonant frequency of the plate.

The experiments could be divided into two main areas. The first set of experiments was devoted to the polarization conductors of the panel. This is the most fundamental element of the design that affects the characteristics of the panel as a whole. For this reason, the greatest emphasis was placed on experimenting with the influence of the drivers on the characteristics of the panel. The second area of experimentation was devoted to the panel's own resonant frequency. Theoretical assumptions suggested that the resonant frequency is defined by the distance of the polarization conductors from the conductive reflecting surface, in our case of the aluminium sheet. The aim of one of the experiments was to determine the dependence of the resonant frequency of the panel on the distance of the conductive surface from the conductor's head. The results of the conducted experiments answered the basic requirements for the design of the depolarization panel.

M. Krchňák et al., *Advancements in Antenna Measurement*, Lecture Notes in Electrical Engineering 1108, https://doi.org/10.1007/978-3-031-48835-1_6

With regard to the individual structural elements, which are the basic pillars of the depolarization panel, the experiments gave us a picture of the interaction of the electromagnetic wave with the parallel system of polarization conductors of the panel.

This made it possible to define recommendations on the thickness of the conductors used, the distance between the conductors, the orientation of the conductors, their arrangement and the distance of the reflecting surface from the conductors, i.e. the thickness of the panel, etc.

The course of the experiments could be divided into the following three basic steps.

1. Design of a measurement device—an experimental panel with the possibility of changing and modifying the main structural elements of the panel. The experimental panel had its own design.
2. Measurement programme design—the experiment required repeated measurements after changing the panel design. It is most convenient to automate the measurement process, which will partially eliminate the influence of the human factor on the measurement results, and the results will be of a consistent nature. The LabView development environment was chosen for measurement automation because of the existing partial automation of the original workstation.
3. The measurement process itself with subsequent evaluation of the results—the measured data were in the form of primary data (tables), which it was desirable to process either graphically or statistically. MATLAB and Excel proved to be the most suitable tools for evaluating large amounts of measurement data.

6.1 Design Solution of the Experimental Reflector Panel

Due to the limited dimensions of the anechoic chamber and the uniformity of the individual measurements, the initial dimensions of the experimental panel (500 $\times$ 500) mm were chosen. This size provided a sufficiently large reflective surface for the measurement antennas used in the GHz frequency range. The use of larger dimensions would partially suppress the effects of multiple reflections of the electromagnetic wave, which would positively affect the results. In this case, however, it was not possible to use much larger dimensions because it was considered to use the original manipulator to fix the experimental panel. The manipulator had structural, dimensional and weight limitations. One of the limitations is the distance of the arm from the floor of the chamber when the arm is raised by 90° from the zero position (horizontal position of the arm). The second limiting factor is the load capacity of the manipulator and the possibility of mechanically attaching the measurement plate to the arm of the manipulator.

The selection of the most suitable drivers for this application was preceded by an extensive survey of suitable and available drivers on the market. The requirements for the drivers were financial, accessibility and, above all, the mechanical and electrical characteristics of the drivers. One of the requirements was the selection of suitable

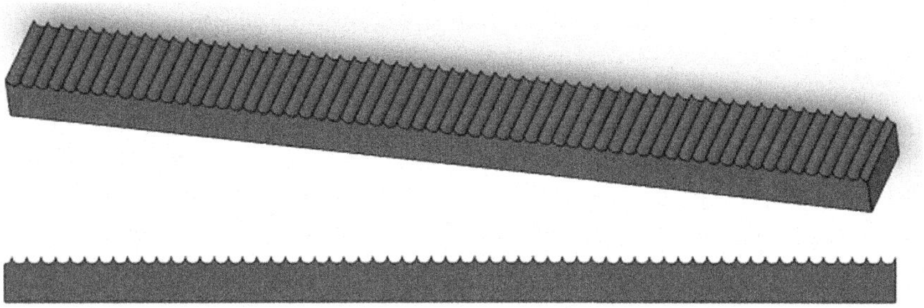

Fig. 6.1 Design solution of the plastic comb

materials for the conductors. For example, steel wires are inexpensive, but their electrical properties are not satisfactory. On the other hand, copper conductors have suitable electrical characteristics, but are more expensive. The solution is to use conductors with a steel core and a copper surface. They are characterized by a low price and their electrical characteristics are close to those of Cu conductors. They are available in diameters {1.6; 2.0; 2.5; 3.2} mm and in lengths of 1 m. Another advantage is that the conductors are not insulated, so the measurement results are not affected by the dielectric properties of the insulation. For this experiment, a conductor diameter of $\varphi = 1.6$ mm was chosen. The wires were cut to the required length of 500 mm. The disadvantage of using single wires is the need to design a technical solution to maintain a constant distance between the wires on the panel.

For this purpose, a PVC "plastic comb" was used (Fig. 6.1). The plastic material is 7 mm thick. Grooves with a diameter of $\varphi = 1.6$ mm and a distance between the grooves of 1 mm were milled into the material. These combs were glued to a supporting polystyrene plate with a thickness of 30 mm. Such a design solution allows to adjust the number of wires and thus to increase the distance between the wires.

A constant distance must be maintained between the conductors and the conductive reflective surface. This is achieved by using extruded polystyrene. See Appendix E, Table 1, for a table of dielectric constants for common materials. Polystyrene has a dielectric constant εr in the range of 2.4–3. From the table, it can be seen that air, beeswax, Teflon, unspecified perfectly dry wood or unspecified rubber have a similar dielectric constant. In terms of mechanical properties, the use of wood or Teflon would be satisfactory. Teflon in the required thickness is financially very expensive and such a plate would cost one hundred euros. The use of perfectly dry wood is practically impossible. Wood absorbs moisture from the environment, which changes its parameters. This would have undesirable effects on the functionality of the experimental panel and we would not be able to guarantee the consistency of the results. A possible technical solution is the use of an air gap. This solution has been used in later experiments. The material labelled Styrofoam is noteworthy. Styrofoam is a trade name for a material made of extruded polystyrene.

It is the most commonly available form of polystyrene, which can be used for thermal insulation of buildings, packaging materials, or food packaging. According to the given table, the dielectric constant is close to 1, which is similar to air or vacuum.

Therefore, it was chosen as the most suitable material. It is a widely available, easily machinable material available in a wide range of thicknesses (from 10 to 200 mm). Another advantage is its low density and non-absorption. The disadvantage is possible evaporation due to UV radiation, which only occurs with long-term exposure to sunlight. The thickness H of the polystyrene for the experimental panel was chosen to be 30 mm. The chosen thickness was determined from the theoretical assumptions in Chap. 4. The panel was designed for a frequency of 2.5 GHz. This frequency was chosen to take into account the capabilities of the laboratory measurement equipment and the attenuation of the coaxial cabling. The theoretical basis indicates that the distance of the conductive surface from the conductors is λ/4 of the required operating frequency.

$$H = \frac{1}{4} \times \frac{c}{f} = \frac{1}{4} \times \frac{3 \times 10^8\,\mathrm{ms}^{-1}}{2.5 \times 10^9\,\mathrm{s}^{-1}} = 0.03\,\mathrm{m} \tag{6.1}$$

where

c is the speed of light in a vacuum;
f is the required panel frequency.

Figure 6.2 shows a 3D model of the construction of an experimental panel with a description of the individual parts. On the basis of this model, an experimental panel was made. The panel also includes a wooden stand that keeps the panel at the required angle of 45°.

6.2 The Effect of the Driver Layout on the Reflective Properties of the Panel

The purpose of the measurement was to experimentally verify the influence of the distance between the conductors on the panel's properties in terms of reflection and resonant frequency. With these measurements, it is possible to prove that the theoretical assumptions made in Chap. 2 are valid. For each distance between the conductors, measurements were made with the following arrangement of elements in the chamber.

Types of measurement:

1. Transmitting antenna with vertical polarization, receiving antenna with vertical polarization, without conductive surface behind the conductors.
2. Transmitting antenna with vertical polarization, receiving antenna with vertical polarization, with a conductive surface behind the conductors.
3. Transmitting antenna with horizontal polarization, receiving antenna with vertical polarization, without conductive surface behind the conductors.

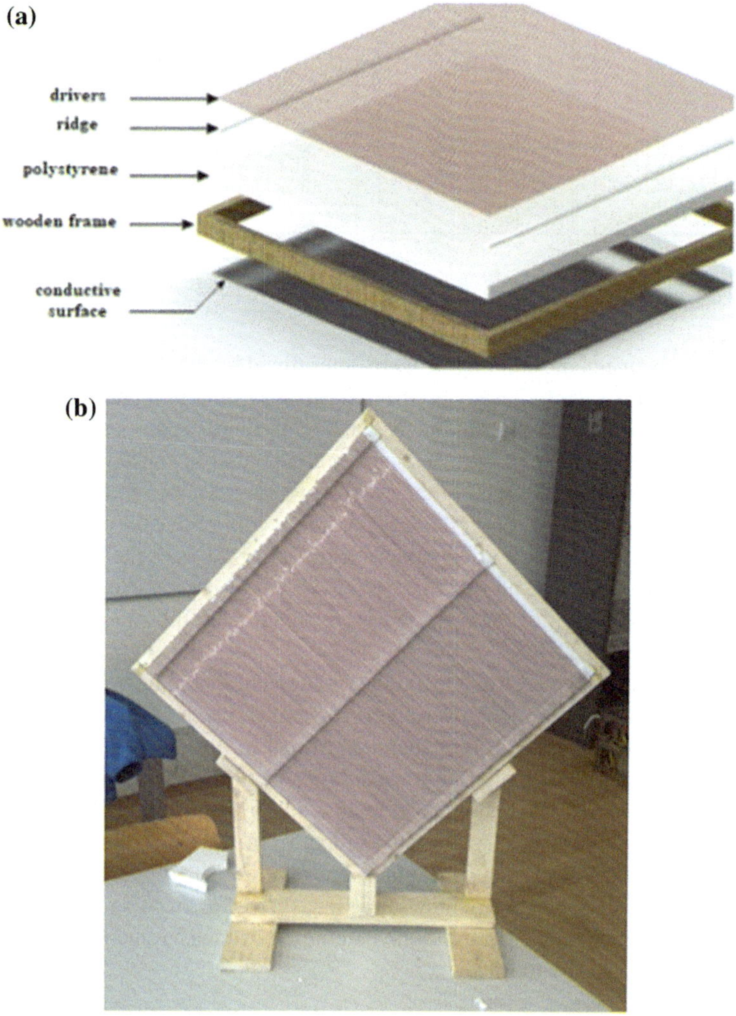

Fig. 6.2 **a** 3D model of the experimental panel; **b** a photograph of the product

4. Transmitting antenna with vertical polarization, receiving antenna with horizontal polarization, with a conductive surface behind the conductors.

The rendering of the above types of measurements, defined by antenna polarization and the use (non-use) of the conductive surface, is illustrated in Fig. 6.3.

The working frequency band was selected in the range of 800 MHz–3 GHz. If necessary, the range was extended to 4 GHz. The step frequency was 250 MHz. The signal level at the output of the generator was 1 V, t.j. 13.01 dBm. Technically, it was not possible to use a signal with a higher level because the signal generator did not allow it.

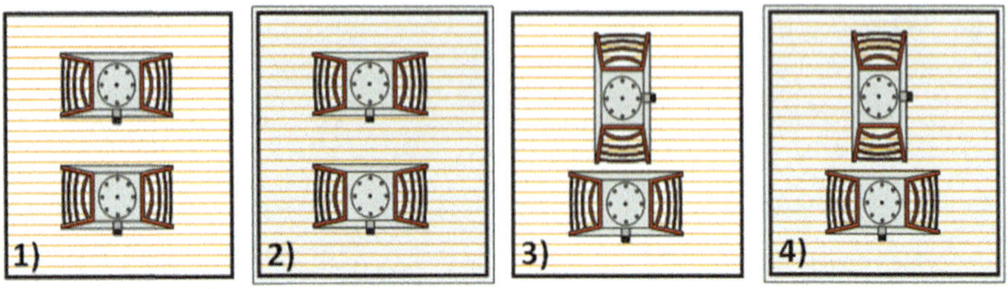

Fig. 6.3 Graphical representations of the types of reflection measurements[1]

The gaps between the conductors were determined by the design of the panel, and their change was made by removing the conductors from the ridge in the order 1; 2; 3; 7; 12 conductors, representing the distance between the centres of the conductors {1; 4; 11; 24; 33} mm. The thickness of the conductors is constant throughout the measurement, as is the distance of the conductive surface from the conductors.

Such a series of measurements should provide a comprehensive picture of how the gap between the conductors affects the reflective properties of the panel. The panel response waveforms were plotted in a MATLAB environment. The results show the dependence of the level of the reflected wave signal from the panel on the angle of rotation of the panel.

The experimental panel was placed in the attenuation chamber on the arm of the manipulator. All measurements were performed at 0° elevation (horizontal position of the manipulator arm) and the panel was filmed in the range of $0° \div 360°$ with two different steps. In the first measurements, a step of 10° was used. Such a step is only suitable for informative measurements. For all measurements, a step of 2.5° was chosen, which provides a sufficient number of measured samples when rotating the panel and a fine rendering of the measured data. At 0° rotation, the wires were oriented horizontally. The frequency range measured was from 1 to 3 GHz.

Figure 6.4 shows an illustrative course obtained by measurement. In the acquired courses, it was possible to trace.

1. The overall shape of the curve, the ideal curve, is without distortions in the area of the minimum and maximum.
2. Signal level values at 45, 90 and 135°.
3. Shifting the maximum or minimum signal level from the corresponding angle.
4. The effect of frequency on the level of the reflected signal.

The predicted waveform is derived from theoretical data. When measuring Type 1, it is expected that at the base position (Fig. 6.3(1)); i.e. the polarization is perpendicular to the conductors, the reflection from the conductors will be minimal, and the maximum reflection will occur when turning by 90°, explained in more detail in Sect. 2.5.

[1] The number in the figure corresponds to the type of measurement.

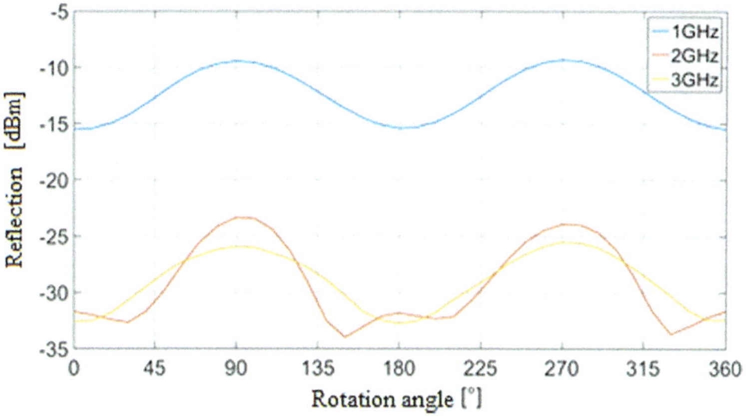

Fig. 6.4 Course of reflection from the panel, measurement Type 1

Adding a sheet, i.e. Type 2 measurement (Fig. 6.3(2)), should increase the level of the reflected signal, but should not affect the shape of the waveform. Measurements of Type 3 (Fig. 6.3(3)); they should show at which rotation the maximum reflection from the drivers occurs. At this value, the polarization of the reflected wave is rotated 90°, i.e., a depolarization effect occurs. Type 4 measurements (Fig. 6.3(4)); aim to verify the effect of the conductive surface on the level of the reflected signal.

Finally, the level of the signal passing through the wires was measured. The analogy of the measurement was similar to that of the previous case, with the difference that the receiving antenna was placed behind the wires on the same axis as the transmitting antenna.

In Fig. 6.4, during reflection from the conductors, four boundaries are indicated at which attenuation values were observed during all measurements. The course shown is a Type 1 measurement (Table 6.1).

The curve in Fig. 6.4 confirms the theoretical assumption. If the polarization of the wave is perpendicular to the direction in which the conductors are arranged, the wave passes through the conductors and only a small part is reflected back, i.e. there is little reflection. If the conductors are parallel to the polarization of the wave, the wave is reflected by the conductors and the maximum reflection occurs. In the range of 45

Table 6.1 Table of attenuation values for the course of reflection

Angle of rotation (°)	Frequency		
	1 GHz	2 GHz	3 GHz
45	– 12.52 dB	– 30.85 dB	– 28.21 dB
90	– 9.44 dB	– 23.34 dB	– 25.92 dB
135	– 12.62 dB	– 30.81 dB	– 28.33 dB
180	– 15.41 dB	– 31.81 dB	– 32.71 dB

and 135°, there is a so-called half bounce. Half the energy of the wave is reflected back from the conductors and half the energy passes through the conductors.

From the measured waveforms shown, the following was determined. The course of the reflection was measured with Type 1 and Type 3 measurements. The endpoint was the maximum and minimum on the reflection path. The maximum signal level is shown on the waveforms. The coordinates determine the signal level and the angle of rotation of the conductors at which the reflection of the wave from the conductors with the corresponding level occurs. The most significant influence of the distance between the conductors on the reflection properties of the panel is observed with the same polarization of the antennas.

After analysing the Type 1 measurements, it can be concluded that the results at a frequency of 1 GHz (Fig. 6.5a) confirm the theoretical assumptions made in Sect. 2.4. The maximum reflection is at 90° and the minimum at 0 and 180°. This phenomenon was confirmed for all the conductor spacings. By analysing the shape of the characteristic, the narrowing of the characteristic is clearly described. For frequencies of 2 GHz (Fig. 6.5b), there is a minimum characteristic of the area around 45°. For a frequency of 1 GHz, the minimum reflection is still in the range of 0 and 180°. Changing the distance between the conductors has no significant effect on this curve. At a frequency of 2 GHz, the reflection curve has a tapered character, to such an extent that the minimum of the overflow is not at perpendicular polarisation, but at an angle of rotation of about 50°. At this frequency, there is a decrease in the reflection characteristic, i.e. a reduction in the angle of rotation of the wires, when maximum and minimum reflection occurs. This phenomenon is probably due to the decomposition of the vectors on the conductors in a different way from that described in the theoretical assumptions, and it has not been possible to qualify it. The characteristic has a greater reflection when the wires are rotated at an angle of 45° than when the wires are rotated at 135° (Fig. 6.6).

At a frequency of 3 GHz, the panel with a large gap of 33, 24 and 11 mm shows approximately the same shape characteristic as at a frequency of 1 GHz. When the distance between the conductors is 1 and 4 mm, the curve of the graph has a tapered character. At smaller spacings, the decay/deformation of the characteristic is shown at frequency and the minimum angle is shifted beyond 45°. The difference between the minimum level at 45 and 135° varied with frequency in the range of 4–5 dBm. From theory, it follows that the decomposition of the wave vectors is not the same when the conductors are rotated to + 45° or − 45°. The emergence of an unbalanced and narrowed characteristic at this frequency is probably due to a non-integral multiple of the wavelength-to-conductor ratio. At a frequency of $f = 2$ GHz, the wavelength $\lambda = 0.15$ m, the length of the conductors is 0.5 m, i.e.:

$$H = \frac{L}{\lambda} = \frac{0.5\,\text{m}}{0.15\,\text{m}} = 3.33 \tag{6.2}$$

In the measured waveforms, the points of minimum reflection were marked and from this value, the values found to the left and right with a decrease of 3 dBm were taken. These values are shown in Table 6.2. The marked points to the left and right

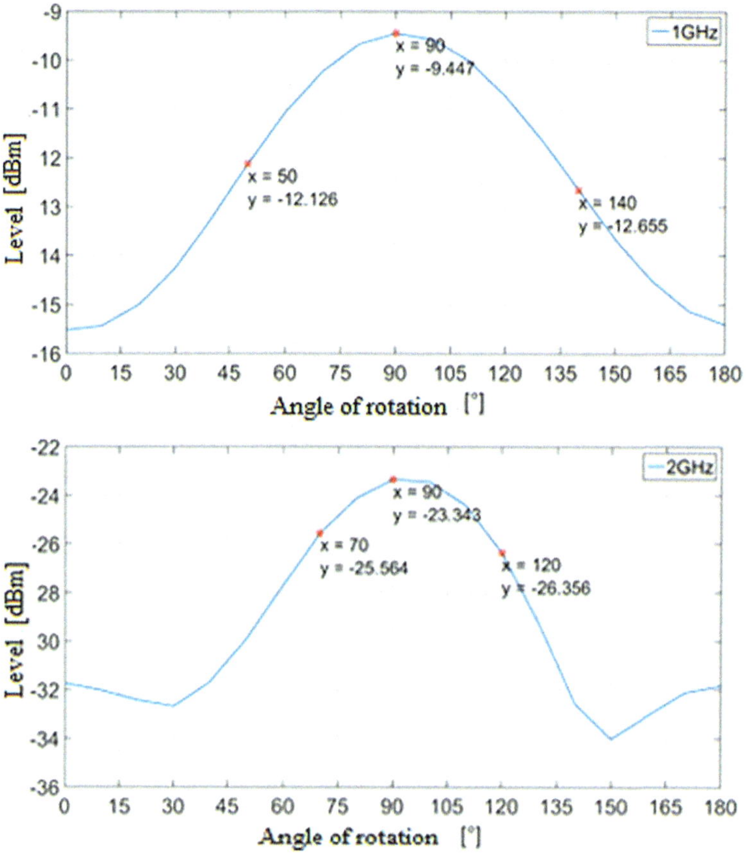

Fig. 6.5 Results of measurement of reflection from drivers

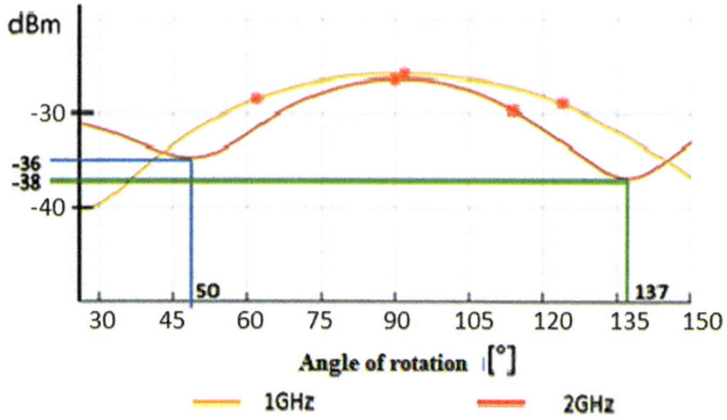

Fig. 6.6 Endpoints for progress

of the minimum value are slightly unbalanced. This is due to the algorithm used to find the corresponding measurement points.

The algorithm subtracted − 3 dBm from the minimum value and then found in the table of measured data the nearest value of the level and its coordinate on the X-axis and marked it on the graph.

The table shows that the angle of rotation of the wires at which the maximum reflection of the signal occurs varies only within a maximum of ± 8°. Changing the distance between the wires does not affect this particular parameter. What is significantly affected is the shape of the characteristic and the width of the characteristic when 3 dBm is dropped. For a distance of 4.11 mm, the width of the curve does not change. There is a slight difference at a distance of 24 mm. At this distance, the trace width increases by 4° at 2 GHz and by 8° at 3 GHz compared to the 11 mm gap and by 12° compared to the 4 mm gap.

A significant change is made by changing the distance between the wires to 33 mm. The increase in the width of the trace is noticeable. At 1 GHz it is 12° for a 24 mm gap, at 2 GHz it is 50° and at 3 GHz it is 22°.

Analysis of the measurements shows that the waveforms are concentrated at certain attenuation levels. Frequencies below 1.2 GHz are concentrated in the region of about − 20 dBm. Below this frequency, there is a drop to a level of about − 30 dB for frequencies from 2 to 2.8 GHz. Higher frequencies than 2.8 GHz are concentrated in the region of about − 40 dBm. This phenomenon was caused by a decrease in the

Table 6.2 Points from the characteristics of wave reflection, the same polarization of the antenna

Space (mm)	Frequency (GHz)	Maximum reflection		Drop − 3 dB on the left		Drop − 3 dB on the right		Width
		(dBm)	(°)	(dBm)	(°)	(dBm)	(°)	(°)
1	1	88.0	− 8.7	− 11.7	51	− 11.7	127	76
	2	94.0	− 26.9	− 29.8	59	− 30.0	124	65
	3	89.0	− 26.1	− 29.0	59	− 29.3	117	58
4	1	90.0	− 7.2	− 10.2	52	− 10.3	130	78
	2	94.0	− 26.2	− 29.1	72	− 29.4	116	44
	3	92.0	− 26.2	− 29.2	64	− 29.4	122	58
11	1	86.0	− 7.2	− 9.9	50	− 10.2	128	78
	2	90.0	− 26.3	− 28.9	70	− 29.6	114	44
	3	92.0	− 25.7	− 28.4	62	− 28.9	124	62
24	1	88.0	− 6.8	− 9.7	52	− 10.1	130	78
	2	92.0	− 27.0	− 30	68	− 30.2	116	48
	3	94.0	− 26.7	− 29.6	58	− 29.9	128	70
33	1	88.0	− 8.5	− 11.4	48	− 11.7	132	84
	2	92.0	− 28.5	− 30.5	32	− 31.7	128	96
	3	92.0	− 29.2	− 32.1	44	− 32.3	136	92

energy of an electromagnetic wave at a fixed distance between the antennas and the depolarisation plate.

Similarly, Table 6.2, which is valid for the same polarisation of the antennas, is similar to Table 6.3, where the values are taken from waveforms at different antenna polarisations. In this case, two basic characteristics of the reflection can be noted. The first is the fact that at a distance between the wires of more than 1 mm and at a frequency of 1 GHz, the minimum reflection of the panel is reached when the wires are rotated to an angle of 135°.

The deformation of the shape of the characteristic compared to the ideal shape occurs due to the ratio of a large gap between the conductors and a small wavelength. This fact was also manifested in the case of measuring the reflection with the same polarization of the antennas. The second feature common to all waveforms with different polarization of antennas is the width of the section where minimum reflection occurs. As in the previous case, the section is monitored with a drop of – 3 dBm. By observing the change in width, it is shown that the change in the distance between the conductors and the change in frequency does not affect the change in the section width in this type of measurement. In all cases, the section width was 50°. Minor deviations could have been caused by the methodology of subtracting section boundaries. It follows that such measurement and evaluation clearly cannot determine the influence of the conductor distance on the reflective properties of the conductor panel. The only indicator of change is the limit value of the distance

Table 6.3 Points on the characteristics of wave reflection, different polarization of antennas

Space (mm)	Frequency (GHz)	Maximum reflection		Drop – 3 dB on the left		Drop– 3 dB on the right		Width
		(dBm)	(°)	(dBm)	(°)	(dBm)	(°)	(°)
1	1	– 20.9	40.0	– 22.6	20	– 24.7	70	50
	2	– 28.0	50.0	– 30.7	20	– 31.5	70	50
	3	– 35.7	50.0	– 38.6	20	– 38.8	70	50
4	1	– 17.9	136.0	– 21.2	108	– 21.2	160	52
	2	– 28.4	50.0	– 31.3	24	– 31.8	72	48
	3	– 30.3	48.0	– 32.9	24	– 33.7	74	50
11	1	– 18.8	136.0	– 21.6	110	– 22.3	160	50
	2	– 29.6	50.0	– 32.4	24	– 33	72	48
	3	– 31.4	50.0	– 33.9	21	– 34.7	74	53
24	1	– 16.9	134.0	– 19.7	110	– 20.3	160	50
	2	– 30.9	48.0	– 33.4	20	– 33.9	70	50
	3	– 32.5	50.0	– 35.9	24	– 35.1	76	52
33	1	– 15.8	44.0	– 18.3	20	– 19.3	70	50
	2	– 31.3	140.0	– 34.2	114	– 34.8	160	46
	3	– 36.4	50.0	– 39.3	24	– 39.6	74	50

of 33 mm between the conductors and the frequency of 3 GHz, when the reflection courses, in the case of the same and different polarization of the antennas, are deformed. This fact shows the inappropriateness of combining a large gap between conductors with a high-frequency electromagnetic signal.

6.3 Influence of the Layout of the Conductors on the Attenuation Properties of the Panel

The second type of measurement was based on the measurement of the signal passing through the grid of wires. The purpose of this type of measurement was to evaluate how the rotation of the wires and the gap between the wires affect the passing signal. The wires were located on the panel, as in the previous case, and formed an obstacle to the signal between the transmitting and receiving antenna. The distance between the transmitting and receiving antennas is 80 cm, with the panel with wires located at the same distance from the antennas, i.e.j. 40 cm. Theoretical assumptions for this measurement are given in Sect. 2.5. These are used to evaluate the results of the experiment. These measurements were made with only one conductor diameter. Structurally, it was impossible to conduct an experiment with a distance between conductors less than 1 mm. In the basic position of the panel, i.e. the wires are horizontal, and the transmitting and receiving antennas have vertical polarization, ideally, the signal should pass through the panel without attenuation. If the direction of the conductors will be identical to the polarization, the signal will not pass through the grid of conductors.

Looking at the results (Fig. 6.7), it is possible to see the confirmation of theoretical assumptions at certain frequencies and a certain distance between the conductors. On the measured waveforms, the effect of the wavelength-to-distance ratio between the conductors can be observed. This effect is most pronounced at a frequency of 1 GHz ($\lambda = 30$ cm). At a minimum distance between the conductors of 1 mm (Fig. 6.7 shown in Green), the waveform at 1 GHz is distorted compared to theoretical assumptions.

It is clearly visible that in the area of 90° rotation of the panel, where the signal should have a minimum level, an area is formed where the level rises and again falls. This creates an area with a slight increase in the level and, accordingly, a second minimum. The first minimum occurs already when the wires are rotated to an angle of 60°, and the signal here reaches a level of – 36.65 dBm, followed by a slight increase in the signal level, where at 90° the rotation reaches a maximum level of – 22 dBm. In the course, a second minimum appears, at a rotation of 110° with a level of – 31.7 dBm.

From this angle, the signal level increases as expected, until the moment when the perpendicular between the conductors and the polarization of the wave occurs, and the passing signal reaches a maximum level – 9 dBm.

By increasing the gap between the conductors to a distance of 4 and 11 mm, the waveform resembles each other. In the field of minimum signal, e.g., at 90° rotation of

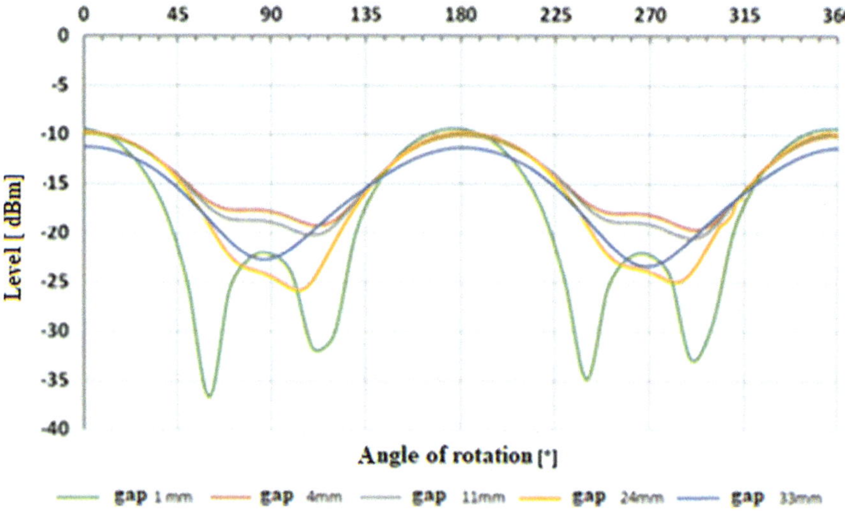

Fig. 6.7 Dependence of the signal level on the gap between the conductors

the panel, there is no longer such a sharp drop and increase in the signal, as in the case of a gap of 1 mm. The electromagnetic wave begins to pass through the conductors according to theoretical assumptions, however, the difference between the maximum and minimum signal levels is about 10 dBm, and the local increase in the signal level in the region of the minimum is still noticeable. The change occurs when the gap between the wires increases again to 24 and 33 mm. At a distance of 24 mm the local increase in the level in the region of the minimum (90°) is practically insignificant and reaches a level of about 1 dB, the characteristic resembles a theoretical assumption. The course at a distance of 33 mm is without deformations, there are no additional level drops and their increase, the level has a periodic course and completely meets the theoretical assumption from the above chapter.

A closer analysis of the measurements revealed a similar effect of the distance between the conductors on the passing signal for other frequencies. The frequencies 1, 2, and 3 GHz were first selected to roughly estimate the behaviour of the electromagnetic wave. Subsequently, 4 additional frequencies of 0.8, 1.2, 2.4 and 2.8 GHz were selected to analyse the point at which the waveform distortion occurs. From these measurements, it is established that the gap between the conductors significantly affects the passage of the electromagnetic wave, and the magnitude of the impact changes significantly with increasing/decreasing frequency. Increasing the gap between the wires had a negative effect on the signal with a higher frequency. At a distance between the conductors of 1 mm is on the course (Fig. 6.8) at frequencies of 2 and 3 GHz the opposite effect can be observed, as described above, for a frequency of 1 GHz. These frequencies have a periodic course, and there is no distortion in the region of the required minimum. The Minimum occurs at an angle

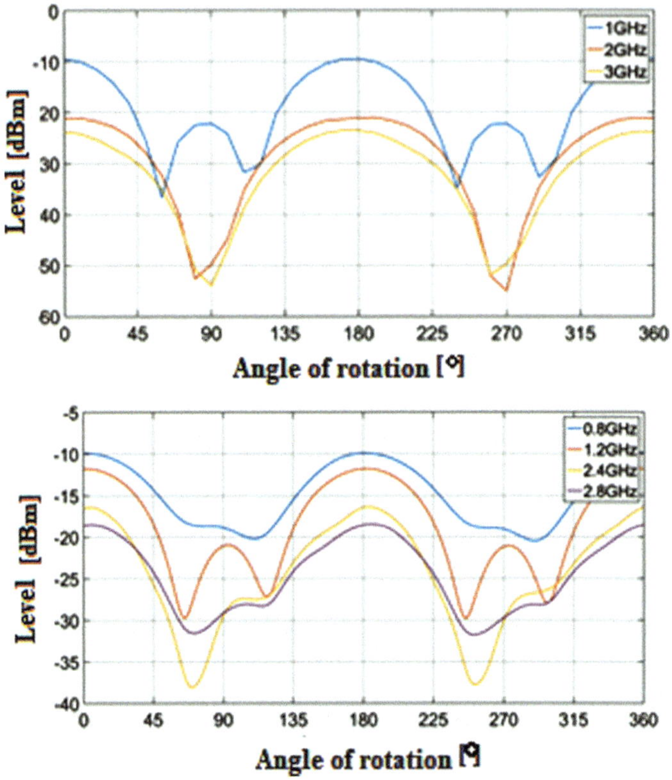

Fig. 6.8 Signal from the parallel arrangement of wires, gap 1 mm

of rotation of 90°, thereby fulfilling the basic theoretical assumption. Deformation can be observed at a frequency of 1 GHz.

Looking at the course (Fig. 6.8) it can be seen that by increasing the frequency, the deformation decreases and gradually disappears, and a region with one minimum occurs. Deformation in the form of a sudden drop in the level creating a local minimum, is found at all combinations of distances between conductors and frequencies, approximately at the same angle of rotation. The local minimum appears when the wires are rotated to an angle of 60–65°. The Maximum occurs when the wires are rotated 90°. The second local minimum occurs at an angle of 110–115°.

Increasing the gap to 4 mm (Fig. 6.9) begins to negatively affect the higher frequencies. The above-mentioned phenomenon of the formation of two overflows occurs in the place where the characteristic should reach a minimum. On the contrary, the waveform at 1 and 0.8 GHz is gradually levelled, and the distortions are lost. This effect is further applied when increasing the gap between the conductors.

At the largest gap of 33 mm (Fig. 6.10) frequencies from 2 GHz have an undefinable waveform resembling white noise. The combination of higher frequencies and a large distance between the wires made the panel not behave according to theoretical

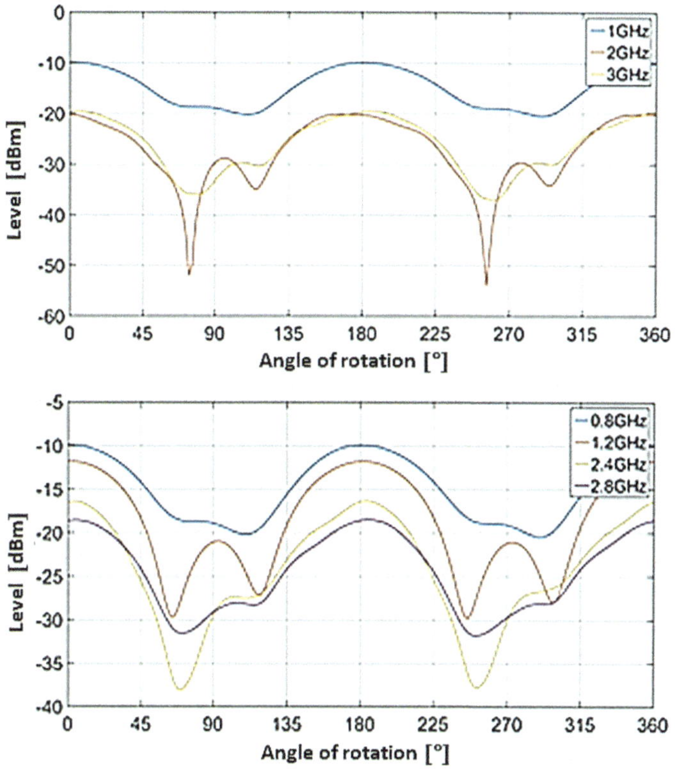

Fig. 6.9 Level behind the parallel arrangement of conductors, gap 4 mm

assumptions. The graphs show that the signal passes through the arrangement of the wires at almost the same level, regardless of the rotation of the wires.

The results obtained are important from the point of view of designing the design of the depolarization panel for different frequencies. The following facts were found, which together form the conclusion of the measurement.

The passage of the signal through the experimental depolarization panel depends on several factors. It depends on the mechanical dimensions of the panel, the number of wires and their length, which further implies the distance between the wires and the applied frequency of the emitted signal towards the panel. The analysis showed that the attenuation depends on the dimension of the conductor relative to the wavelength.

The conductor dimension may be an integral multiple or fraction of the wavelength of the transmitted signal, or an increment $\lambda/4$ or $\lambda 3/4$. The conductors themselves, under the influence of the electromagnetic field, behave like antenna elements. Electromagnetic energy is received and re-emitted. It is believed that the deformation of the waveform is due to the fact that the electromagnetic wave emitted by the antenna does not fall with an identical phase on all conductors. In the direction from the centre to the edges of the conductors, the elements are phase-delayed, at the same

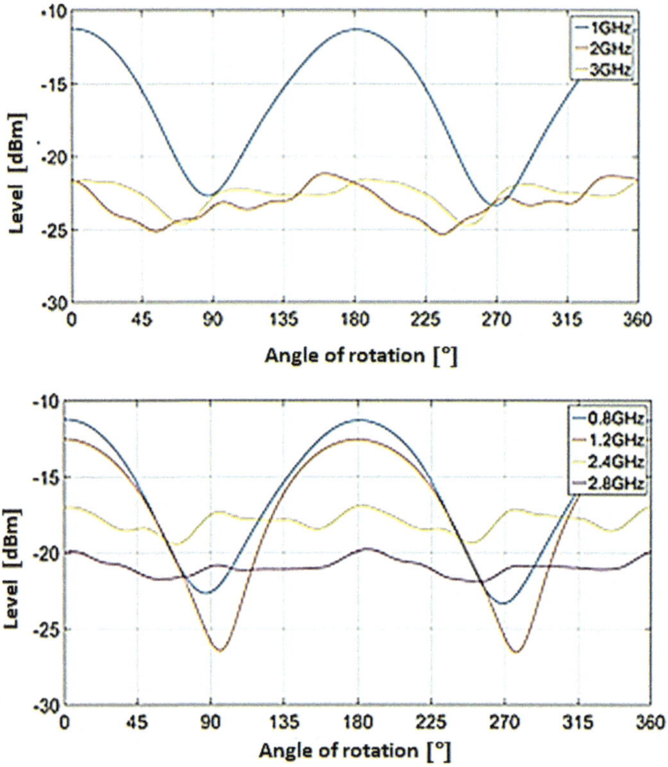

Fig. 6.10 Level behind the parallel arrangement of conductors, gap 33 mm

time, when each element radiates energy again, it radiates again with a phase delay, but with the phase rotated by 180°. The resulting wave, leaving the conductors, is composed of different, phase-delayed signals, and at the point of reception of the signal creates waveform distortions. If the energy falls on the entire surface of the conductors with the same phase, there is no phase delay and distortions.

The graphs showed that when constructing a panel for frequencies below 1 GHz; it is advisable to use a gap between the wires of more than 33 mm. On the contrary, for frequencies greater than 1 GHz, a suitable gap is as small as possible. For this reason, a Conrad (50 × 0.035) mm^2 tape conductor was used in the construction of additional depolarization panels intended for testing.

6.4 Working Frequency of Depolarization Panel

The basic parameter that defines the properties of the depolarization panel is its operating frequency. This frequency is determined by the mechanical design of the panel. Proceeding from the basics of Chap. 2, it becomes clear that the working area of the panel, that is, its operating frequency, will be very narrow. Thus, the Panel is narrow-band. Since the panel uses the principles of composing an electromagnetic wave based on phase delays and the ratios of the incident and emitted signal, there is no known way to create such a panel for a wide frequency spectrum. In the design of the panel, emphasis must be placed on the exact observance of the thickness of the dielectric within the entire surface of the panel. As described in the above chapter, the thickness of the dielectric must be exactly $\lambda/4$ of the wavelength to which the panel needs to be tuned. The design of the experiments described in this chapter is based on theoretical assumptions about how the depolarization panel should react to an electromagnetic wave.

The construction of depolarization panels intended for experiments took place in several steps. The first step was to determine the required operating frequency. The operating frequency was chosen for the 2.4 GHz region, that is, for the lower frequency of the Wi-Fi connection. This is the most widespread frequency, with the help of which almost all modern communication devices create communication. The use of Wi-Fi in aviation is already practically commonplace, and such a connection is offered by all modern airlines.

To calculate the required thickness of the depolarization panel, a simple relation was used:

$$f = \frac{c}{4h} \tag{6.3}$$

$$h = \frac{c}{\frac{f}{4}} = \frac{3 \times 10^8}{\frac{2.4 \times 10^9}{4}} = 0.03125\,\mathrm{m} \tag{6.4}$$

where

c is the speed of light in a vacuum;
f is the required operating frequency;
h is the thickness of the panel in meters.

In Table 6.4, the selected frequencies and calculated dielectric thicknesses of the depolarization panel are given. From the table, it is clear at first glance that in the manufacture of a depolarization panel, it is very important to emphasize compliance with the thickness. It can be seen that at some frequencies, accuracy to hundredths of a millimetre is required. Any deviation from this distance re-tuned the frequency, depending on whether the deviation is positive or negative in relation to the calculated thickness value.

Extruded STYRODUR polystyrene is used as a dielectric in the manufacture of depolarisation plates for experiments. This material is produced in thicknesses of (20;

Table 6.4 Dielectric properties of materials

Material	ε_{rmin}	$\varepsilon_{r\ min}$	Material	$\varepsilon_{r\ min}$	ε_{rmin}
Air	1.0	1.0	Nylon	3.4	22.4
Amber	2.6	2.7	Paper	1.5	3
Asbestos fibres	3.1	4.8	Paraffin	2.0	3.0
Bakelite	5.0	22	Plexiglass	2.6	3.5
Beeswax	2.4	2.8	Polycarbonate	2.9	3.2
Batist	4.0	4.0	Polyethylene	2.5	2.5
Celluloid	4.0	4.0	Polyamide	3.4	3.5
Acetate fibre	2.9	4.5	Polystyrene	2.4	3.0
Durite glass	4.7	5.1	Porcelain	5.0	6.5
Ebonite	2.7	2.7	Quartz	5.0	5.0
Epoxy resin	3.4	3.7	Rubber	2.0	4.0
Fibre	5.0	5.0	Muscovite	5.4	5.4
Umakart	3.6	6.0	Selenium	6.0	6.0
Glass	3.8	14.5	Shellac	2.9	3.9
Glass pyrex	4.6	5.0	Silicone	3.2	4.7
Gutta-percase	2.4	2.6	Slate	7.0	7.0
Isolantite	6.1	6.1	Dry soil	2.4	2.9
Kevlar	3.5	4.5	Talc shale	5.2	6.3
Polymethyl methacrylate	2.5	2.5	Styrofoam	1.03	1.03
Mica	4.0	9.0	Teflon	2.1	2.1
Composite micarta	3.2	5.5	Vaseline	2.16	2.16
Mycalex	7.3	9.3	Vinylite	2.7	7.5
Neoprene	4.0	6.7	Distilled water	34	78
Dry wood	1.4	2.9			

30; 40; 50; 60; 80; 100; 120; 140; 160; 180 and 200) mm. Therefore, the calculated thickness values have been rounded to these values. For the operating frequency of 2.4 GHz, according to Table 6.5 requires a thickness of 3.13 cm, the nearest available value is 30 mm. This theoretically shifts the operating frequency to 2.49 GHz, which is an acceptable shift for experiments.

In the course (Fig. 6.11) it is seen that the working frequency decreases with the power of the thickness of the dielectric according to:

$$f_p = 75h_d^{-1}\ \text{(GHz, mm)} \tag{6.5}$$

where

f_p is the operating frequency (GHz);
h_d is the thickness of the dielectric (mm).

Table 6.5 Thickness of depolarization panel by frequency

Required frequency (GHz)	λ (cm)	Dielectric thickness (cm)
1	30.00	7.50
1.2	25.00	6.25
1.4	21.43	5.36
1.6	18.75	4.69
1.8	16.67	4.17
2.0	15.00	3.75
2.2	13.64	3.41
2.4	12.50	3.13
2.6	11.54	2.88
2.8	10.71	2.68
3.0	10.00	2.50
3.2	9.38	2.34
3.4	8.82	2.21
3.6	8.33	2.08
3.8	7.89	1.97
4.0	7.50	1.88

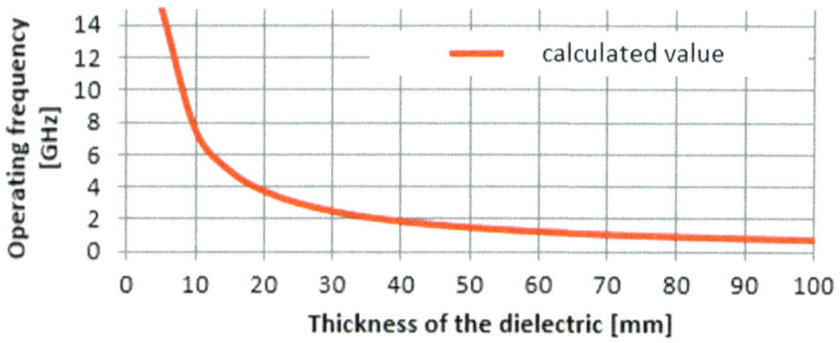

Fig. 6.11 Theoretical dependences of the working frequency on the thickness of the dielectric

In this way, it is possible to easily design the thickness of the dielectric according to the required frequency, or, conversely, determine the frequency according to the available thickness of the dielectric.

6.5 Experimental Determination of Working Panel Frequencies

Two methods were proposed to determine the operating frequency, which was derived directly from the measured results. Each of the methods was aimed at tracking a different parameter from the measured results. The result of the measurements is a detailed specification of the frequency characteristics of each panel produced. The conclusions drawn from these measurements make it possible to give recommendations for the design and manufacture of depolarisation panels for specific requirements and to predict the characteristics of the proposed panels. The measurement methodology itself can be applied to the manufactured panels to obtain a detailed description.

- Method 1—Monitoring the angle of rotation of the wires at which the maximum and minimum reflection of the signal occurs. This method is experimental and time-consuming. It does not directly show the frequency at which the panel is tuned but only helps to understand the complex properties of the panel.
- Method 2—Measuring the frequency response of the panel with the same and different polarisations of the reference antennas. This method directly determines the operating frequency of the depolarised panel. The method is fast and simple and can be performed directly at the measurement site for verification purposes.

The objects of the measurement depolarisation panels are shown in Fig. 6.12. A series of small square depolarisation panels differ in the thickness of the dielectric and the arrangement of the conductors.

Legend to Fig. 6.12:

- Panel No. 1 has a diameter of 100 cm, with a dielectric thickness of 3 cm, f = 2.5 GHz.
- Panel No. 2 has a diameter of 50 cm with a dielectric thickness of 3 cm, f = 2.5 GHz.
- Panel No. 3 has a thickness of 4 cm and the conductors are of the same length, dimension 50 cm × 50 cm, f = 1.8 GHz.
- Panel No. 4 has a dielectric thickness of 3 cm, with conductors directed parallel to the edge dimension 50 cm × 50 cm, f = 2.5 GHz.
- Panel No. 5 has an arrangement of conductors diagonally at an angle of 45° and a thickness of 3 cm, dimensions 50 cm × 50 cm, f = 2.5 GHz.
- Panel No. 6 has a dielectric 2.5 cm thick, the conductors are parallel to the edge dimension 50 cm × 50 cm, f = 3 GHz.
- Panel No. 7 contains a 2 cm dielectric, the conductors are parallel to the edge dimension 50 cm × 50 cm, f = 3.7 GHz.
- Panel No. 8 dielectric 3 cm thick, conductors, diagonal, size 120 cm × 120 cm, f = 2.5 GHz.

Fig. 6.12 Implemented types of depolarization panels

6.5.1 *Angular Method of Determining the Operating Frequency*

The measurement methodology for Method 1 was similar to that described in Sect. 5. 2. An illustration of the antenna layout is given in Fig. 6.3(2), (3).

When measuring the reflection characteristics of the parallel array of conductors, it was observed that the measured waveforms changed shape with increasing frequency and concentrated at certain levels. This phenomenon is described in more detail in Sect. 5.2.

A panel thickness of 3 cm was chosen as a reference. Figure 6.13 shows how the reflection curve of the panel changes as the frequency of the electromagnetic wave increases. Looking at the reflection curve in the frequency range from 1 to 1.4 GHz, it can be seen that the waveforms in this range have a deformed shape compared to Fig. 6.4, the origin of which has already been described. The angle of rotation of the

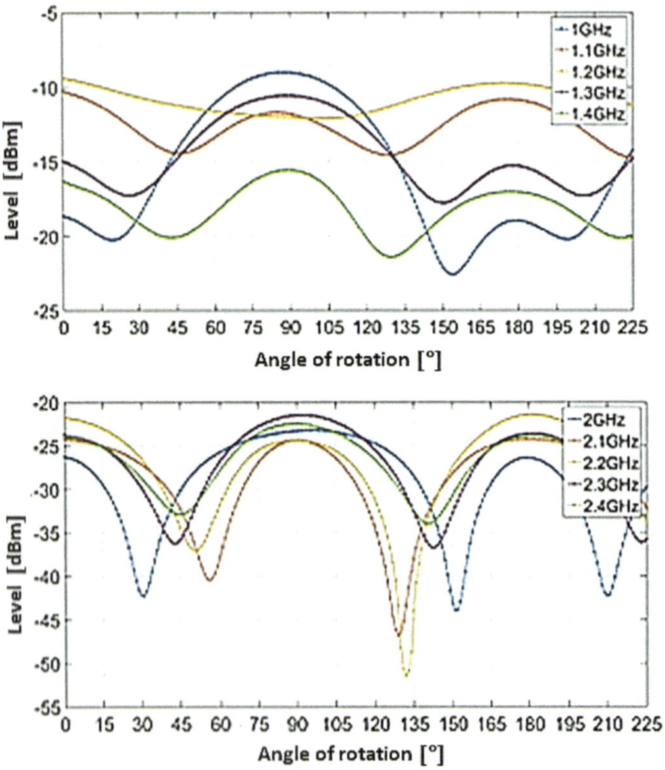

Fig. 6.13 Level of the reflected wave from depolarization panel 1–2.4 GHz

wires at which the maximum reflection occurs is 89–90° at a frequency of 1 GHz, and the same applies at frequencies of 1.3 and 1.4 GHz.

However, the minimum reflection does not occur at an angle of 45°, but at 20° for 1 GHz and 29° for 1.3 GHz. The minimum reflection at 45° only occurs at 1.4 and 1.1 GHz. There is an anomaly at 1.2 GHz. The course of the reflection at this frequency is inverted, i.e., the minimum occurs at an angle of 90° and the maximum at angles of 0 and 180°. The origin of this anomaly could not be clarified. Figure 6.13 shows that the reflection curve does not change significantly up to 1.9 GHz. There are no sharp decreases towards the local minimum and no sharp increases towards the local maximum of the reflection. From this behaviour, it can be concluded that the panel is outside its operating frequency range. At frequencies above 2 GHz, the shape of the response curve changes, which manifests itself in the limits from 2 to 2.8 GHz. In this range, the working frequency is theoretically calculated. The curve begins to show sharp dips and sudden outbreaks of local peaks and troughs. The local minimum is concentrated in the area of the 45° rotation of the wires. The local maximum of the reflection occurs in this frequency range in all cases at an angle of rotation of 90°. By observing the coordinates of the local minimum and maximum,

Table 6.6 The emergence of a local minimum reflection

Frequency (GHz)	Angle of rotation (°)
2	30
2.1	58
2.2	47
2.3	44
2.4	45
2.5	59
2.6	53
2.7	30
2.8	60
2.9	60
3.0	50

it is possible to estimate the operating frequency by comparing the measured angle of rotation at which the minimum occurs with an angle of 45°. Similarly, for the maximum and 90°.

The rotation angles at which the local minimum is formed are summarised in Table 6.6. The table is valid for a dielectric thickness of 3 cm and the same polarisation of the antennas. From the data, it can be seen that when the signal frequency is brought closer to the working frequency of the panel, the local minimum shifts to the 45° region. This angle is reached at a frequency of 2.4 GHz. Above this frequency, there is a shift away from the 45° angle. As the frequency is increased above 3 GHz (Fig. 6.15), the characteristics begin to deform and show a shape that is difficult to define, making it possible to estimate the operating frequency to which the panel is tuned. A similar analysis was carried out for panels 2 cm thick and 4 cm wide.

A similar reflection behaviour as described for a 3 cm thick panel is also observed for a 2 cm thick depolarization panel (Table 6.7). In this case, the theoretical operating frequency of the panel is in the region of 3.6 GHz. From 1 to 3 GHz there is a distortion of the characteristics. This indicates that the panel does not behave ideally in this frequency range and that the operating frequency is not here. When the frequency limit of 3 GHz is exceeded up to 3.9 GHz, a gradual alignment of the characteristics towards a similar shape can be observed. By observing the local minimum, it can be seen that the minimum reflection at a 45° conductor angle occurs at a frequency of 3.4 GHz and not at a frequency of 3.6 GHz as calculated. This frequency shift can be caused by a number of factors such as uneven dielectric thickness or waviness of the conductive substrate. Table 6.7 presents the evaluation of the working frequency of panels by Method 1.

Waveforms with different antenna polarisations show similar behaviour as described in Sect. 5.2. Analysis of the trace in Fig. 6.14 shows that the waveforms for reflection with different antenna polarisations do not change as much as for reflection with the same antenna polarisation.

Table 6.7 Evaluation of the working frequency of panels by Method 1

Dielectric thickness (cm)					
2		3		4	
Frequency (GHz)	Angle of rotation (°)	Frequency (GHz)	Angle of rotation (°)	Frequency (GHz)	Angle of rotation (°)
2.9	47	2	30	1	29
3	44	2.1	58	1.1	43
3.1	31	2.2	47	1.2	–
3.2	44	2.3	44	1.3	31
3.3	42	2.4	45	1.4	33
3.4	45	2.5	59	1.5	46
3.5	33	2.6	53	1.6	44
3.6	40	2.7	30	1.7	40
3.7	44	2.8	60	1.8	42
3.8	32	2.9	60	1.9	58
3.9	42	3	50	2	28

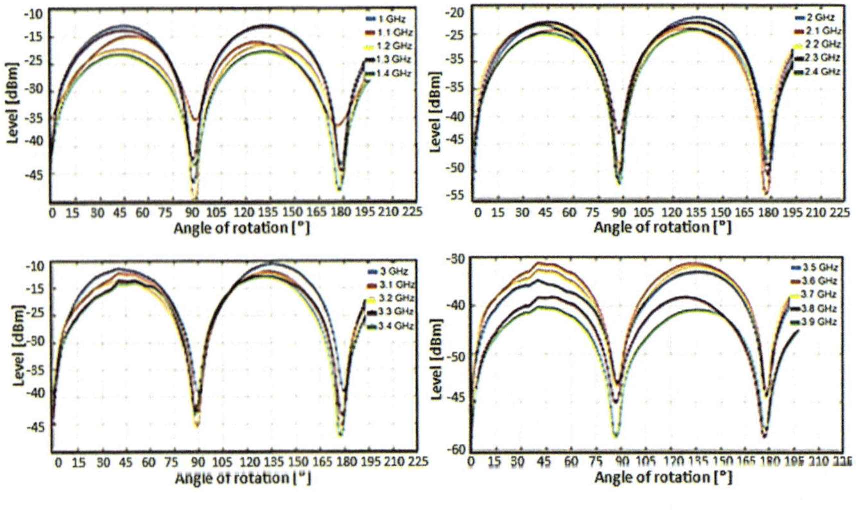

Fig. 6.14 Level of reflection, different polarization of antennas

The waveforms do not show significant distortions outside the working area of the panel. In these measurements, the angle at which the local minimum occurs was observed, but the reference value was not 45° but 90°.

The local maximum should occur at 45°. The local minimum is found at all frequencies in the range of 89–91°. The local maximum is in the range of 44–47°. The waveforms have a similar shape and nothing distinguishes them to the extent

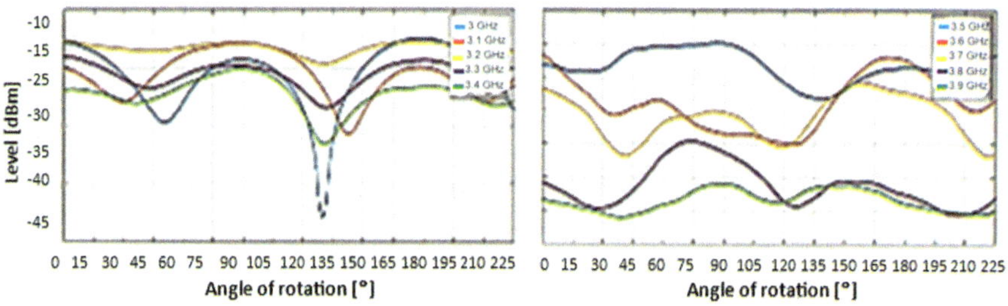

Fig. 6.15 The level of the reflected wave from the depolarization panel 3–3.9 GHz

that it is possible to select the operating frequency unambiguously. The situation is slightly different with a dielectric thickness of 4 cm, where distortions in the characteristics are observed at frequencies of 2.5 GHz and above. The calculated working frequency is 1.8 GHz. Up to 2 GHz, the waveforms show the same shape and again, it is not possible to determine the operating frequency unambiguously. However, it can be said that in the region above 2 GHz, this operating range will not exist. In the case of a 2 cm thick dielectric, where the operating frequency has been calculated to be 3.6 GHz, no waveform distortions are visible across the frequency spectrum. However, it can be seen that at low frequencies, the waveforms split into levels. This phenomenon has already been described in Chap. 2, and it is not possible to unambiguously determine the working frequency of the panel from measurements with different antenna polarization, as was the case for measurements with the same antenna polarisation (Fig. 6.15).

The graphical evaluation of this methodological approach is based on the plotting of graphs from tables similar to Table 6.7. For this evaluation, an algorithm has been developed in MATLAB that searches for local minima and maxima of waveforms and stores them in the form of CSV for further processing.

The working range of the panel can be determined by a graphical representation of the angles at which the local minimum or maximum occurs. Observing the course of the minimum, it can be seen that up to 3 GHz, the course has large deviations from the ideal value (the course has a large variance). From the frequency of 3–4 GHz, the deviation decreases and a region appears where points appear when the peaks of the characteristic are at 45°. Above the frequency of 4 GHz, the characteristic oscillation reappears. From this behaviour and previous analyses, it is possible to estimate that the working range of the panel will be in this area (marked on the graph by a black dashed line). Looking at the curve of the maximum reflection value, it is not possible to clearly define such a boundary. The maximum reflection still occurs in the region of 90 and 180°, with considerable scattering at very high frequencies, above 4 GHz. This difference between monitoring the local minimum and the maximum is due to the fact that at the local minimum, the signal level is monitored with very low energy $(-30 \div -50)$ dBm, so even small deviations in the rotation of the wires will be reflected there significantly. The signal energy at the maximum reflection is in the

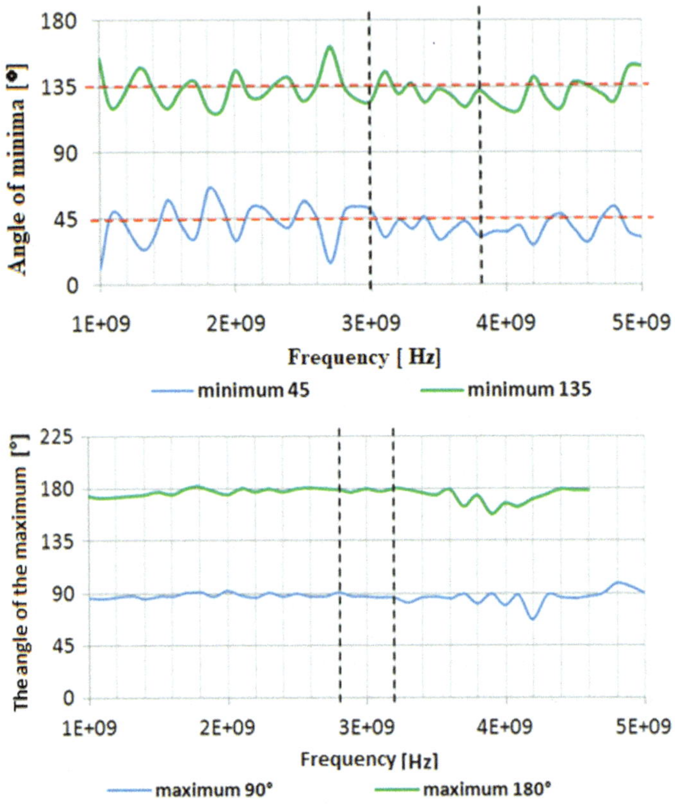

Fig. 6.16 Working frequency band of panel H = 2 cm determined by Method 1[2]

range (− 10 ÷ − 25) dBm, where it does not manifest itself to such an extent. Similar curves have been made for 3 and 4 cm panels (Fig. 6.16).

It was again confirmed that monitoring the maximum reflection curve is not very useful, as the corresponding characteristics in Figs. 6.17 and 6.18 do not show a sufficiently large dispersion to identify the working area. However, it has been confirmed that monitoring the local minimum can determine the boundaries of the frequencies at which the panel will operate. These limits are shown in Figs. 6.17 and 6.18 with a black line. In the case of a 3 cm panel, the region of lowest dispersion from 2.1 to 2.8 GHz is visible. Below and above this limit, the dispersion of the graph increases. The same is true for a 4 cm panel. The operating limit determined by this method is in the range of 1.2–2 GHz when monitoring the local minimum. The first method gives an estimate of the frequency range in which the panel will operate. However, this method is very demanding in terms of technical equipment, time and processing of the results. It is therefore not very suitable for frequent repetition.

[2] Panel with a dielectric thickness of 2 cm and the same polarization of antennas.

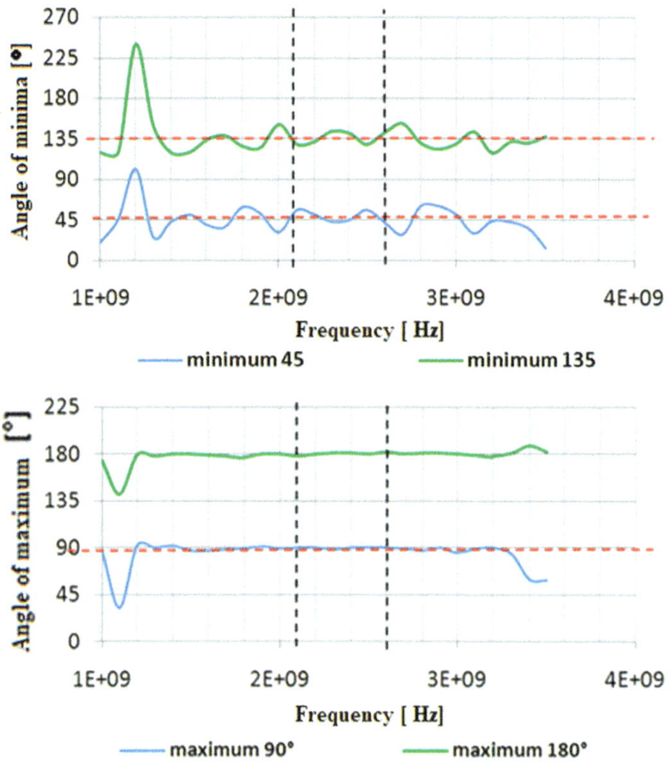

Fig. 6.17 Working frequency band of panel H = 3 cm determined by Method 1[3]

6.5.2 Frequency Method of Determining the Operating Frequency

Method 1 was found to be suitable for initial experimental measurements to determine the basic frequency characteristics of the panel. Due to its complexity, a simpler Method 2 has been proposed. Method 2 is based on stationary measurements of the reflection of the electromagnetic wave from the depolarisation panel itself. This means that it is not necessary to move the depolarisation panel or the antenna system significantly. While in Method 1 the antennas are fixed during the measurement, but the depolarisation panel rotates in the range of 360°, in Method 2 both the antennas and the depolarisation panel are in a fixed position. At the same time, the orientation of the panel wires is at an angle of 45° or 135° relative to the polarisation of the antennas. This method of measurement allows the frequency characteristics of the panel to be analysed directly in the laboratory where the measurement is carried out.

[3] Panel with a dielectric thickness of 3 cm and the same antenna polarization.

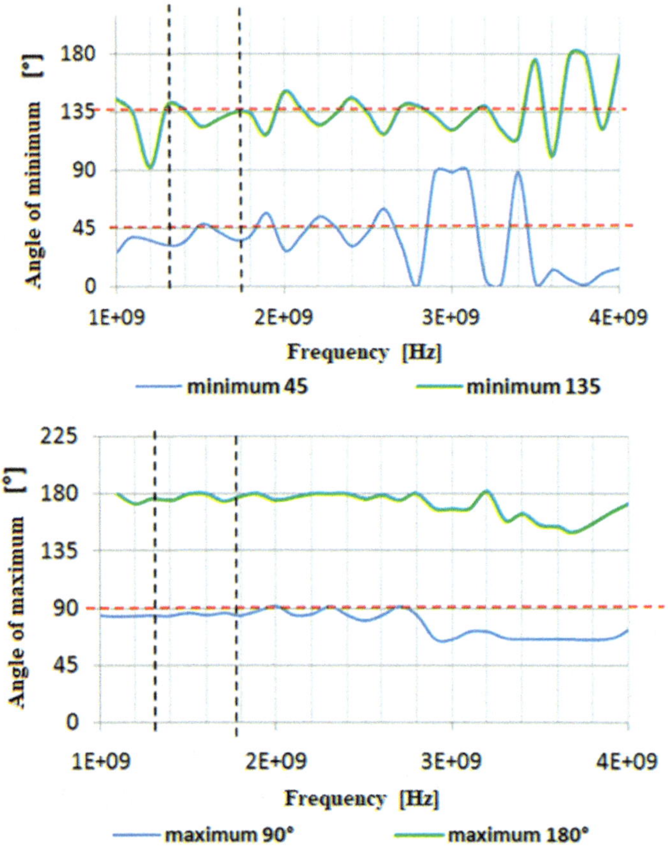

Fig. 6.18 Working frequency band of panel H = 4 cm determined by Method 1[4]

The arrangement of the antennas, their polarisation and the orientation of the panel conductors for Method 2 are shown in Fig. 6.19. Figure 6.19a, b show the positions of the antennas and the oblique orientation of the conductors of the square panel at an angle of 45° compared to the panels in Fig. 6.3. Figure 6.19c, d are photographs of the position of the antennas and the oblique orientation of the wires of a real circular panel at an angle of 45° on a real measurement workstation. The measurement is therefore carried out with two different antenna positions. In the first case, both antennas (transmitting and receiving) have the same polarisation (Fig. 6.19a, c). In the second case, both antennas (transmitting and receiving) have a different—transverse—polarisation (Fig. 6.19b, d).

When measured according to Fig. 6.19c, if both antennas (transmitting and receiving) are equally polarised and at an angle of 45° to the depolarisation field, the transmitted signal should be reflected from the ideal depolarisation field at a minimum

[4] Panel with a dielectric thickness of 4 cm and the same polarization of antennas.

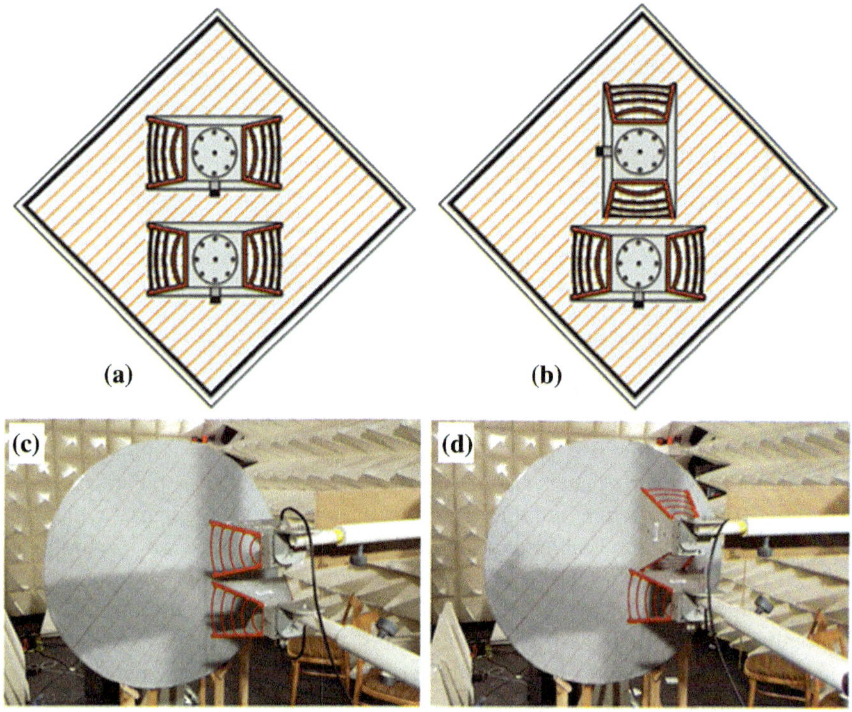

Fig. 6.19 Measurement of the frequency characteristics of a circular panel

level. This is approximately the level at which the signal is reflected from the internal attenuation walls of the attenuation chamber. This is due to the change in polarisation of the transmitted signal after reflection from the depolarisation panel. Since the receiving antenna has the same polarisation as the transmitting antenna, but the reflected signal from the depolarisation panel already has a transverse-perpendicular polarisation to the transmitted signal, the received signal cannot be processed by the receiving antenna. In this case, the cross-attenuation between the transmitting and receiving antennas becomes apparent during reception, mimicking the attenuation of the walls of the attenuation chamber. When measured according to Fig. 6.19d, if both antennas (transmitting and receiving) have perpendicular polarisation to each other and their polarisations are at an angle of 45° relative to the depolarisation panel, the transmitted signal should be reflected from the ideal depolarisation panel at the maximum level. This is approximately the level at which the signal is reflected from a conductive metal plate. This is due to the change in polarisation of the transmitted signal after reflection from the depolarisation panel. Since the receiving antenna also has a different perpendicular polarisation to the transmitting antenna, and the reflected signal from the depolarisation plate also has a transverse perpendicular polarisation to the transmitted signal, the received signal can be properly processed by the receiving antenna. In this case, there is no cross-attenuation between the

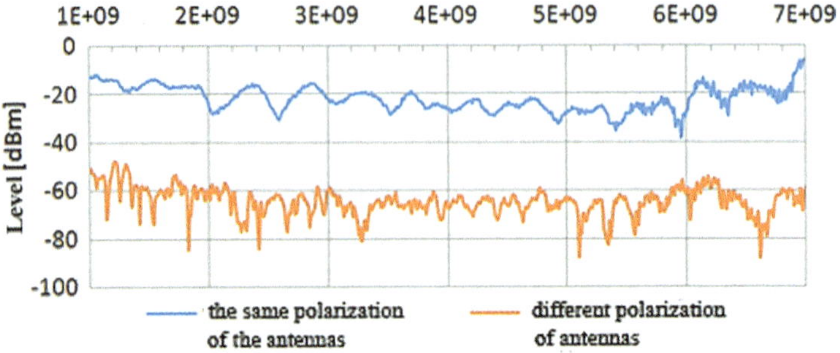

Fig. 6.20 Reflection of the signal from a plate of square shape

transmitting and receiving antennas during reception, mimicking the reflection from a conductive plate (Fig. 6.20).

In fact, depolarising panels are not ideal. They cannot change polarisation over the entire frequency range, but only over the narrow frequency band to which they are tuned. It is also impossible to ensure that the electromagnetic wave is uniformly applied to the entire surface of the depolarising panel, to avoid disturbing phase delays, which manifest themselves as sharp drops in the frequency response. On the real frequency characteristics of the panels, the signal level is expected to fall exactly at the frequency to which the panel is tuned if the measurements are made with the same polarisation of the antennas. For measurements with different antenna polarisations, the level of the reflected signal at the tuned frequency should be as high as possible. The simplest, quickest and most visually appealing way of doing this is to compare the characteristics of the same and different polarised antennas. This gives a peak at the tuned frequency of the panel. The working area is defined as the area where the greatest difference between the same and different polarisation measurements is found and where this area is equalised. Equilibrium refers to a section of a characteristic where the character of the curve does not change, i.e. there are no sudden drops in the surrounding level or the whole characteristic does not show a gradual decrease to another centre of the level. This gradual decrease is caused by signal attenuation in the wiring unless calibration for attenuation is performed. It is important to note that the characteristic of the reflection at different polarisations in the working area of the panel must not be below the level of the reflection measured at the same polarisation of the antennas. When a characteristic is measured over a sufficiently wide frequency range, harmonic resonance frequencies will appear.

First, the reflection from the conductive surface—the plate—is measured. The latter is part of the depolarisation panel and forms its rear wall. The reflection was measured from both the round and square plates (Fig. 6.21).

The simplest, quickest and most visually appealing way of doing this is to compare the characteristics of the same and different polarised antennas. This gives a peak at the tuned frequency of the panel. The working area is defined by the area where the greatest difference between the same and different polarisation measurements is

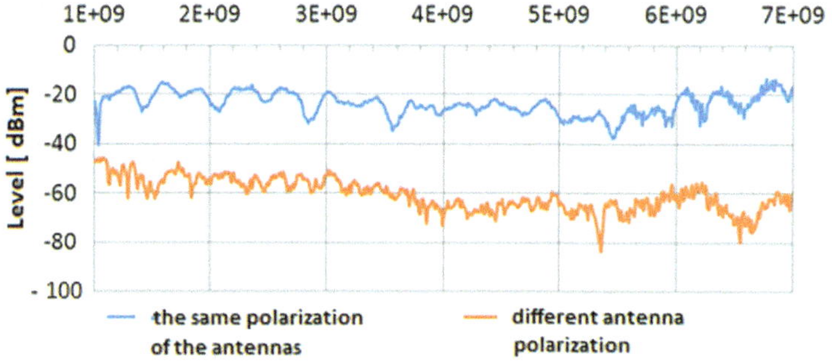

Fig. 6.21 Signal reflection from a plate of circular shape

found and where this area is equalised. Equilibrium refers to a section of a characteristic where the character of the curve does not change, i.e. there are no sudden drops in the surrounding level or the whole characteristic does not show a gradual decrease to another centre of the level. This gradual decrease is caused by signal attenuation in the wiring, unless calibration for attenuation is performed. It is important to note that the characteristic of the reflection at different polarisations in the working area of the panel must not be below the level of the reflection measured at the same polarisation of the antennas. When a characteristic is measured over a sufficiently wide frequency range, harmonic resonance frequencies will appear.

First, the reflection from the conductive surface—the plate—is measured. The latter is part of the depolarisation panel and forms its rear wall. The reflection was measured from both the round and square plates (Fig. 6.22).

When the characteristic is measured with different antenna polarisations, the characteristic shows the same character as when measured with the same polarisation. However, the threshold at which the waveforms begin to resemble each other is shifted from 3 to 4 GHz. Above this limit, the waveforms are identical. The only difference is the height of the two peaks at 5.1 and 6.6 GHz. In the case of a square plate, they have levels of − 87.9 dBm at these frequencies, while for a round plate, they are − 67 and − 79 dBm.

However, up to 4 GHz, the difference between the patterns is quite significant. The reflection pattern of the circular plate does not show any level drop and moves around the limit of − 57 dBm ± 5 dBm. The characteristics are balanced.

The square plate reflection curve shows significant and frequent dips in the signal level. These dips reach values as low as − 85 dBm, while the overall curve fluctuates around − 65 dBm.

By comparing and analysing these two measurements, it was found that the shape of the material affects its reflective properties. A possible explanation for this phenomenon is partially confirmed by the simulations in Sect. 2.8. As in the case of the simulated conductors, the sheet metal has similar properties, i.e. the maximum of radiated energy is at the sharp edges and corners of the object. This secondary radiation can cause significant overflows due to different phase ratios. In the case of a circular plate, there are no sharp edges, i.e. the secondary radiation is not as

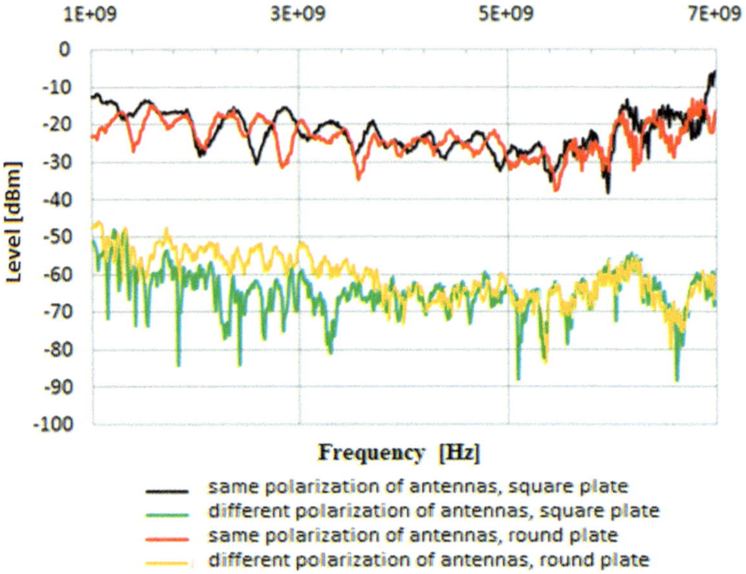

Fig. 6.22 Comparison of reflection from plates of different shapes

pronounced. It can be assumed that in the case of a square sheet, the radiated wave does not fall evenly and simultaneously on the entire surface. The intensity of the incident wave in the centre of the sheet will be higher than at the extreme positions of the sheet, and it will also fall on the edges of the sheet with a phase delay. It is reflected from the surface with the same delay. The use of a circular plate partially eliminates this problem, since from a geometrical point of view there are not so significant differences in the distances from the centre of the surface to the edges.

The characteristic curve (as well as all the others that will be described) shows three areas that are related to the operating frequency. The characteristic curve (as well as all the others that will be described) shows three areas that are related to the operating frequency. The characteristic curve (as well as all the others that will be described) shows three areas that are related to the operating frequency (Fig. 6.23).

These have been defined:

- Working area—in this area, there is a working frequency of the depolarization panel. The area is defined by boundaries centred around the calculated working frequency. The lower limit is defined by the initial frequency and the drop in signal level with different antenna polarization.
- Half-wave resonance area—this area begins by analogy where the work area ends. The area is characterized by the fact that there is a significant level drop in it, measured with different antenna polarization. The nature of the course with the same polarization does not change. This overflow has a value of twice the working frequency. This suggests that the panel is excited with a half-wave resonance.

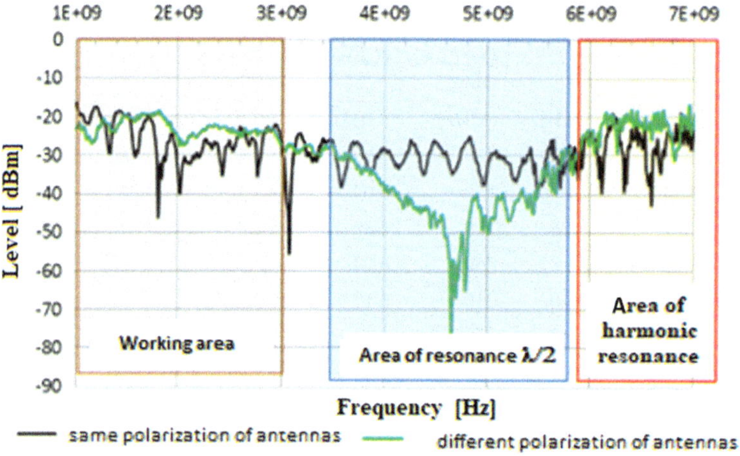

Fig. 6.23 Display of the working areas of the depolarization panel

- Harmonic resonance area—the harmonic resonance area may or may not be present during measurement. If the measurement is completed at a moment behind the half-wave resonance area, the area will not be present. The depolarization panel here has a similar character to the working area but operates at frequency multiples.

It is advisable to measure the characteristics over a wide frequency range in order to correctly identify the operating frequency and operating range. In the case of half-wave resonance, the dip has a certain width, which is significantly different for a circular and a square panel. The analysis showed that this width can be influenced by the angle of incidence of the wave on the depolarization plate. The angle of incidence changes from the centre of the panel to the edge, considering that the funnel antennas used have a conical radiation characteristic. A different angle of incidence of the wave on the depolarizing plate causes a change in the distance the wave has to travel from the conductors to the plate. Since the frequency characteristics of the depolarisation panel are precisely defined by the thickness of the dielectric, changing the angle of incidence of the wave causes a frequency detuning. In the case of a square panel, there will be a greater number of spurious reflections compared to a circular panel because the circular panel has an even distance from the centre of the panel to the edges. With a square panel, this distance from the centre changes, with the greatest deviations towards the corners.

In the region of half-wave resonance, overflow widths are summarized in Table 6.8. Figure 6.12 indicates the panel numbers. Values were obtained by subtracting from the characteristics. Panels number 1 and 2 are circular in shape, and display low overflow widths in this area. Only a marginal effect (a difference of 30 MHz) on the width was observed due to differences in panel size.

Table 6.8 Table of measured operating frequency values overflow widths

Panel number	Working f_p	$\lambda/2$ f	Overflow width	Harmonic f	Theoretical f_t	$(f_t - f_p)$	Level at f_p
	(GHz)	(GHz)	(MHz)	(GHz)	(GHz)	(MHz)	(dBm)
1	2.44	4.62	130	6.02	2.5	60	12.6
2	2.44	4.86	160	6.04	2.5	60	21.6
3	1.87	3.6	210	4.68	1.8	70	19.8
4	2.44	4.85	250	6.04	2.5	60	8.6
5	2.45	5.02	210	6.13	2.5	50	14.1
6	3.1	6.2	260	6.3	3	100	24.8
7	3.68	6.58	–	–	3.7	70	19.2
8	2.47	4.71	430	6.01	2.5	30	12.5

The width varies between 210 and 260 MHz for 50 cm square panels. Panel number 8, which has an edge of 120 cm, shows a significant increase in width. The width is twice that of a 50 cm panel, rounded to the nearest value. In one experiment, the wires were short-circuited by the rear conductive surface. Short-circuiting the conductors led to equalization of the characteristic and suppression of the depolarization phenomenon. A significant manifestation occurs at the operating frequency of 2.44 GHz, where the difference in level is 10 dBm for short-circuited wires and 21 dBm for disconnected wires, resulting in a two-fold improvement in the depolarization phenomenon. No further investigation was conducted for the panel with short-circuited wires. It was not feasible to identify the overflow and the higher harmonic frequency for Panel No. 7. Panels that depolarize at an operating frequency of 2.4 GHz have a thickness of 3 cm. The frequency range of the half-wave resonance is from 2.6 to 4.8 GHz. The frequency did not appear at an exact integer double of the resonant frequency, but at a multiple of 1.9–2.1 times. This minor deviation was anticipated due to measurements being carried out on panels with varied shapes, sizes, and lengths of wires. The sixth panel, with an angular shape and a dielectric thickness of 2.5 cm, exhibited the largest difference in working frequency compared to the calculated frequency. The variation was up to 100 MHz for this panel, whereas for others, the difference ranged between 50 and 70 MHz. The frequency at which half-wave resonance occurs has also shifted to 6.2 GHz. The region of higher harmonic frequencies couldn't be detected due to the previously stated limitations of the instruments. Between panels 4 and 5, there was a difference in wire arrangement between the 4th and 5th panels. It was believed that wires of varying lengths would improve the panel's frequency characteristics and reduce the impact of conductor resonance. Upon comparing the characteristics of panels 4 and 5, it was determined that the panel with diagonal conductor arrangement exhibited opposite properties to those expected, resulting in an impaired frequency response. This is demonstrated by an increase in both the number and size of level drops. The feature exhibits greater oscillation compared to a panel with a uniform conductor arrangement. However, the waveforms have similar characteristics, with only a 10 MHz difference in the operating frequencies of both panels.

The difference between the widths of the overflows was 40 MHz. A panel with the same arrangement of wires had a smaller width. The smallest difference between the calculated and measured operating frequency was Panel No. 8. The depolarisation phenomenon is characterised by the difference between the signal level with different and the same polarisation of the antennas at the operating frequency. In this respect, Panels No. 6 and 2 show the most pronounced depolarisation phenomenon. It reached values of 21 and 24 dBm. The least pronounced depolarisation phenomenon was observed in Panel No. 4, where it reached values of only 8 dBm. Panels 1 and 8 showed a similar level of depolarisation phenomenon at 12.5 dBm. The characteristics of both panels are comparable in terms of the parameters studied. A comparable depolarisation phenomenon was also observed in Panels No. 3 and 7.

6.6 Experimentally Authentication Frequency Characteristics Panel

The measurement was slightly different than the one described in the previous chapter. The measurement methodology is based on Method 2. The measured object in this case was a laboratory preparation that allowed a gradual change in the distance of the conductive surface from the conductors with a step of 5 mm in the range of 0–100 mm. Due to the fact that the frequency range of the measuring apparatus is in the range of 1–7 GHz, results outside this frequency limits have not been evaluated. The theoretical operating frequencies were determined by the graph in Fig. 6.11, according to which the basic operating frequency is defined by the thickness of the dielectric. The measuring laboratory preparation was designed to imitate the properties of the depolarization panel as faithfully as possible. Minor deviations are expected as the design of the fixture is not as precise as the depolarization panel itself.

In the construction of the laboratory setup, the wires were placed on 2 mm thick PVC plastic. Since the plastic itself has a different dielectric constant than extruded polystyrene, it could affect the measurement results and cause deviations in the measured values of the operating frequencies. Based on real measurement results, the original Fig. 6.11 was supplemented with the measured values of the operating frequencies, which after this modification is shown as Fig. 6.24. On the basis of individual measurements (Table 6.9), with the same shape as in the case of measurements with a depolarization panel.

The measured results in the observed frequency range confirm the theoretical calculation. The curve of the measured value of the working frequency of the laboratory preparation has a comparable shape to the calculated working frequency. However, the analysis of the results revealed deviations between the calculated and measured values. This is more pronounced at the highest measured frequencies ($4 \div$

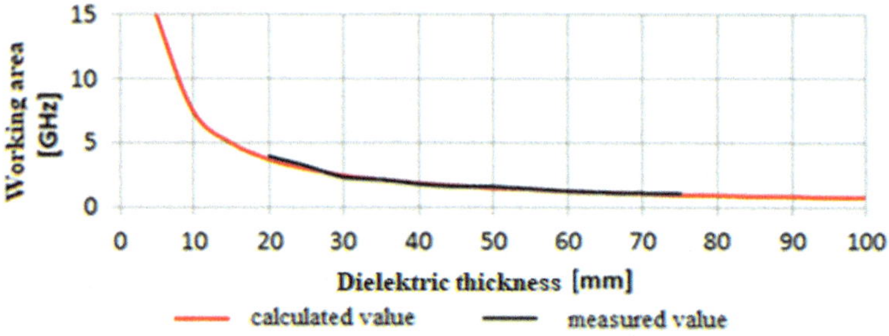

Fig. 6.24 Measured from the operating frequency depends on the thickness of the panel

Table 6.9 Table of measured operating individual measurements frequency values

Distance	Working f_p	$\lambda/2f$	Overflow width	Harmonic f	Theoretical f_t	$(f_t - f_p)$	Level at f_p
(mm)	(GHz)	(GHz)	(MHz)	(GHz)	(GHz)	(MHz)	(dBm)
20	3.89	7.3	130	–	3.75	140	17
25	3.16	5.09	40	–	3.00	160	15
30	2.33	4.68	230	6.3	2.50	170	14
35	2.11	4.13	310	4.94	2.14	30	14.5
40	1.81	3.3	270	4.5	1.88	70	18.7
45	1.6	3.06	170	4.36	1.67	70	19.5
50	1.57	2.86	320	3.8	1.50	70	24
55	1.38	2.61	280	3.6	1.36	20	23.3
60	1.24	2.44	80	3.29	1.25	10	21.5
65	1.1	2.2	140	3.24	1.15	50	17.5
70	1.09	2.09	10	3.21	1.07	20	21.4
75	1.08	2.03	120	2.72	1.00	80	19.7

3) GHz, where it is necessary to create the smallest distances between the conductive surface and the conductors. It is precisely the imperfection of the fixture design that can cause deviations in the measurement.

Looking at the difference between the actual and calculated frequencies, it can be seen that this difference is more pronounced at high frequencies than at lower frequencies. Even a small change in the distance of the conductive surface from the conductors will cause a significant change in frequency (see Fig. 6.24). The most noticeable difference in frequency is at a distance of 30 mm and is up to 170 MHz. The smallest difference is at a distance of 60 mm, where the difference is only 10 MHz. In other cases, the frequency difference ranged from 30 to 80 MHz. These values are comparable to the depolarization panel measurements. The most pronounced depolarization phenomenon occurred at distances of 55 and 60 mm, at levels of 21.53 dB and 23.3 dBm.

The depolarization phenomenon at the remaining distances was around 19 dBm, with a more significant deviation at a distance of 30 mm, where it decreased to 14 dBm. Since the largest deviation between the calculated and measured frequencies was found at this distance, this is probably a measurement error. The panel with a thickness of 70 mm showed the smallest overflow width in the half-wave resonance, where it reached values of 10 MHz. The remaining overflow widths were comparable to those measured on the depolarization panels. For larger distances, from 60 to 75 mm, the overflow width was less than 200 MHz. The results clearly show that depolarization panels have more consistent properties at lower frequencies and greater dielectric thicknesses than at high frequencies (Fig. 6.25).

Fig. 6.25 Preparation for measuring the frequency dependence of the panel

References

1. Labun, J., Krchňák, M., Kurdel, P., Češkovič, M.: Systém rádiofrekvenčnej depolarizačnej komory, patentová prihláška SK 56-2018, A3, 6 s. ÚPV SR, Banská Bystrica (2017)
2. Labun, J., Kurdel, P., Adamčík, F., Češkovič, M.: Odrazový panel na zmenu polarizácie rádiových vĺn, zverejnená patentová prihláška SK 53-2014, A3, 7 s. ÚPV SR, Banská Bystrica (2017)
3. Labun, J., Kurdel, P., Adamčík, F., Češkovič, M.: Systém zníženia účinku parazitných odrazov rádiovýškomera, patentová prihláška SK 54-2014, A3, 6 s. ÚPV SR, Banská Bystrica (2017)

Chapter 7
Optimization Layout Elements Measurement

This chapter is dedicated to one of the practical applications of the depolarization panel. The main intended use of the depolarization panel is to measure the radiation characteristics of antennas. The previous chapter provided important knowledge about the properties of the depolarization panel itself. Thanks to this knowledge, such as working frequency and suitable panel shape, it was possible to design a basic methodology for measuring radiation characteristics using a depolarization panel.

The basic measurement in this chapter is the radiation characteristic measurement of a symmetrical dipole. The purpose of the measurement is to determine the reference characteristic, which will be used later for comparison with the measured results. The reference characteristic is measured in a standardized way—for direct visibility of antennas, without the use of a depolarization panel. This symmetrical dipole was then used as the main object of radiation pattern measurement. The main objective of the measurements in this chapter was to find the most suitable placement of antennas/manipulators for measurements with a depolarization panel. This is mainly the orientation of the measurement and manipulator antennas relative to the panel and to each other, and the distance of the antennas from the depolarization panel.

The second part of the chapter is devoted to measurements of the directional characteristics of reflector antennas designed for the Wi-Fi 2.4 GHz band. These are antennas from SANO Antennas, SANO MICRO 8, SANO SECTOR V16 and an unknown type of reflector antenna with 4 radiators [1, 2]. For measurements with a depolarization panel, the SANO MICRO 8 type was chosen because of its simplest design and simple radiation characteristics. All measurements with these antennas were performed using the laboratory setup, which proved to be the most convenient among the symmetrical dipole measurements. The purpose of the measurements was to verify the functionality of the depolarization panel and the measurement methodology also for directional characteristics.

The third part of the chapter contains measurements to verify the methodology and functionality of the depolarization panel outside the attenuation chamber.

M. Krchňák et al., *Advancements in Antenna Measurement*, Lecture Notes in Electrical Engineering 1108, https://doi.org/10.1007/978-3-031-48835-1_7

The measurements were made outside the attenuation chamber in the laboratory, where there are undesirable reflections of electromagnetic waves from the walls and various obstacles. The results obtained in this section are only a supplement to prove the possibilities of the measurement methodology using the depolarization panel in laboratory conditions.

Due to the necessity of controlling the manipulators and measuring instruments by means of a PC, it was not possible to perform these measurements outside the laboratory room where the attenuation chamber is located.

7.1 Evaluation of Directional Characteristics of Symmetric Dipole

The main object of measurement was an antenna of the symmetrical dipole type. This antenna was chosen because of its radiation characteristic. The characteristic has the shape of an eight and contains sharp minima. Distortions of the measured characteristic are clearly recognizable, and there is no need for a detailed analysis of the results to determine whether an error occurred in the measurement process (e.g. a wrong joint) or whether the measurement was successful. The change in the phase relationships of the signal reflected from the depolarization plate will be manifested by a change in the region of the characteristic minimum. The peaks will not be as sharp, but they will be flattened, with the shape of the characteristic curve deviating negatively from the ideal eight. On the contrary, sharp drops in the minimum area of the measured characteristic indicate that the reflection of the signal from the panel is as good as possible for the application. Obtaining the radiation characteristic of the dipole by reflection from the depolarization panel was a process in which it was necessary to resolve several basic issues (the distance of the panel from the antennas, the layout of the antennas) that fully affect the measurement result. In this section, the measurement results are analysed [3].

For this purpose, a symmetrical dipole was modelled and simulations of the 3D spatial radiation pattern of Fig. 7.1 left and the plane section in polar coordinates of Fig. 7.1 right were performed.

When evaluating the measurements, the result of the reference measurement of the antenna characteristic of the symmetrical dipole (Fig. 7.4) is comparable to the simulation characteristic of a similar dipole (Fig. 7.1). In the case of the antenna characteristic created by simulation, the minimum signal level is obtained when the antenna is rotated to an angle of 0 and 180°.

The effort of the authors of the paper was to observe the position of the characteristic minimum at the same values of 0 and 180° also during the real measurement. The minimum level of the measured characteristic is not exactly at the angles of 0 and 180° but is shifted by about 8° to higher values, i.e. 8 and 188°. A smaller difference could not be achieved mechanically with this method of mounting the antenna on the manipulator.

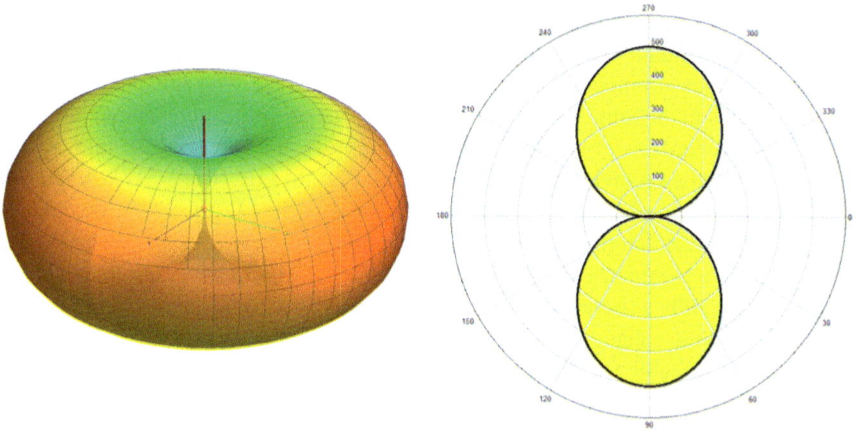

Fig. 7.1 Simulation of the characteristics of the symmetrical dipole

The first series of measurements were carried out with the antenna positioned so that at the beginning of the coordinate system—at zero—there was a maximum of the emitted signal. This caused differences in the signal level measured at 0 and 360°. These values should be identical, but in real measurements with such an arrangement, they turned out to be slightly different. For an example of measurement, frequency 2.4 GHz, $\varphi = 0°$ of dipole rotation, the received signal level is 4.695 mV, but at $\varphi = 360°$ it is 4.987 mV. The difference is only 0.292 mV, but on the characteristic, it causes a deformation in the form of a level jump between these rotation angles.

To meet the need of attaching a symmetrical dipole to the manipulator, a holder for this antenna was designed and then printed using 3D printing (Fig. 7.2). The antenna was glued into the antenna holder, which is part of the manipulator.

The coaxial feed to the antenna was fixed to the manipulator to partially suppress the effect of cable movement when the antenna rotates on the manipulator. The first measurements performed on all antennas tested were VSWR and radiation pattern measurements. We consider these radiation characteristics as a reference since they

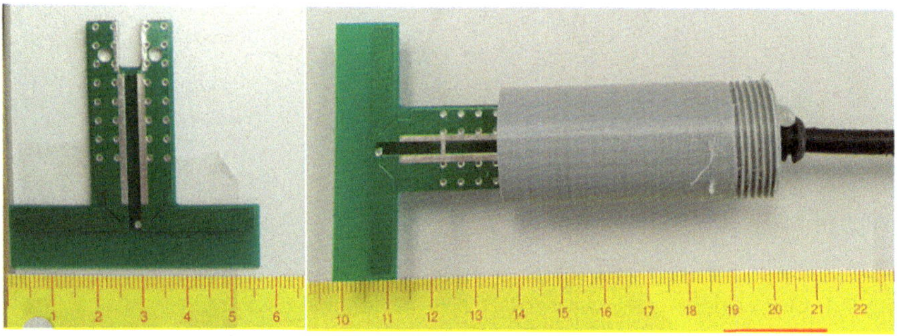

Fig. 7.2 Symmetrical dipole was used as a standard for measuring

were obtained in a standard way as described in Sect. 3.4. The measurement was made with the antennas facing each other, with the distance between the antennas such that the measurement was made in a far field. The receiving antenna was rotated around its axis in the range of 0–360°, while the transmitting antenna was stationary. In all cases, regardless of the layout of the laboratory, the transmitting antenna was stationary and the receiving antenna rotated around its axis. The VSWR waveform shows the matching of the load and source impedances. In the case of antennas, this parameter can determine the operating frequencies of the antenna. The VSWR waveform for a symmetrical dipole is shown in Fig. 7.3. On the graph, the working range is marked by a dashed rectangle. The maximum VSWR value considered is $\delta = 2$. This VSWR value covers the frequency range from 2 to 2.7 GHz. Therefore, the antenna is suitable for use in the selected Wi-Fi 2.4 GHz band.

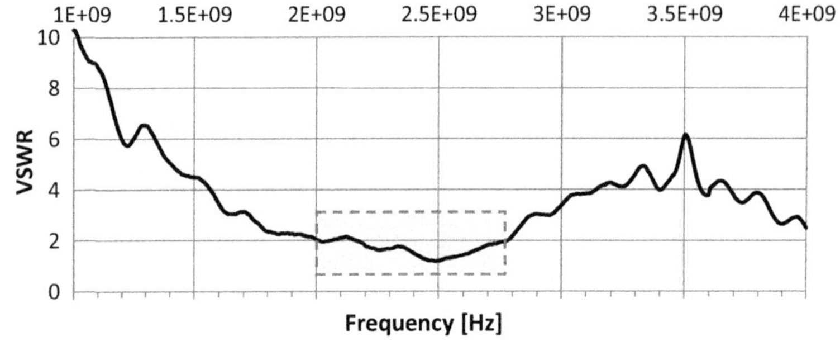

Fig. 7.3 VSWR of the dipole under test

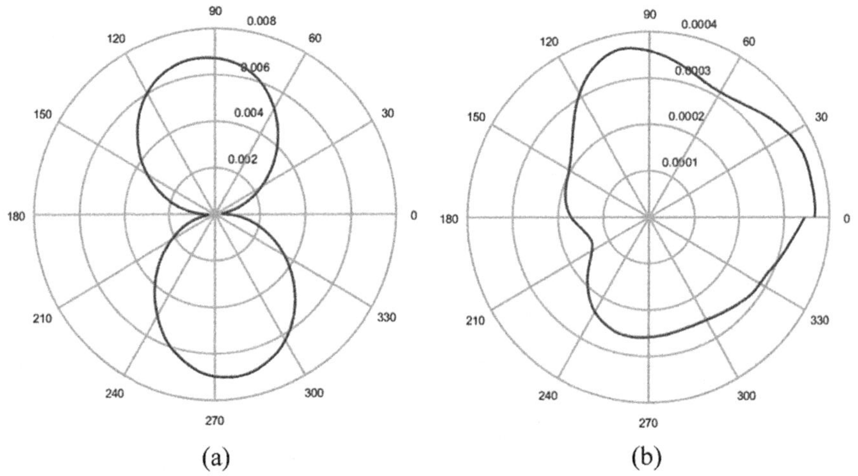

Fig. 7.4 Characteristics of the symmetrical dipole

Once these reference values were obtained, the radiation characteristic of the dipole was measured by reflection from the chamber wall (Fig. 7.4b). This provided evidence that all other measured characteristics were not reflected in the chamber wall.

The emission characteristic is measured in a standard manner and is considered a reference, as shown in Fig. 7.4a. A clear difference in the shape of the characteristic can be observed when compared to Fig. 7.4b. It was impossible to obtain the figure-of-eight-shaped reflection characteristic when measuring from the depolarization chamber wall. This outcome was expected since the chamber walls reflect minimal energy back to the source, as discussed in Sect. 5.4. This measurement eliminates the possibility that the measured characteristics were caused by reflection from the walls of the chamber.

Four depolarization panels were used for the measurement. The panels were selected based on their frequency characteristics, as shown in Fig. 6.12 (Panels 1, 2, 5, and 8). The ETS LINDGREN MODEL 3115 funnel antenna was used as the transmitting antenna.

The measured characteristics exhibit both shape and symmetry similarities. Since the measurements were taken using depolarization panels with similar properties, no diametric differences were expected. The antenna was positioned in the manipulator in a way that minimized the radiation at a 0° angle of rotation. As demonstrated in the previous chapter, square-shaped Panels No. 5 and No. 8 exhibit worse characteristics than round-shaped Panels No. 2 and No. 1, respectively. Therefore, it was expected that measurements taken on square panels would indicate slight distortions. In the first case, the measurement was taken through reflection from Depolarization Panel No. 5. The characteristic has the shape of an eight. The maxima and minima of the antenna radiation are clearly distinguishable. However, there is a certain degree of deformation in the characteristic. The upper lobe has a maximum value of 3.8 mV, and the lower lobe has a maximum value of 3 mV. Replacing the panel with Panel No. 2, which was round with a diameter of 50 cm, resulted in a significant improvement. Comparison with the reference characteristic shows a slight decrease in the level across the measured range, as expected, due to the depolarization panel's characteristic attenuation at a given frequency. The characteristic has the same shape as the reference one. The areas of minimum and maximum emission are clearly definable. A similar result was obtained when using the depolarization Panel No. 1. The signal level is comparable to that of the previous measurement. Nevertheless, the upper lobe displays some amount of deformation in comparison to the lower lobe.

Both measurements on circular Plates No. 1 and 2 are considered correct. In both cases, the characteristics are comparable to the reference value. A deterioration in the shape of the curve occurred when the depolarisation plate was replaced by Plate No. 8. The measured curve is not completely symmetrical. The lower radiation lobe shows a distortion in the form of a slight elongation. The radiation minima do not have such a sharp and clear transition as in the case of Panel 2 and the reference characteristics. The signal level fluctuates within 0.004 mV in all cases. The exception is the measurement with Panel #8, where the signal level dropped below 0.0015 mV. This panel shows significant attenuation compared to the other

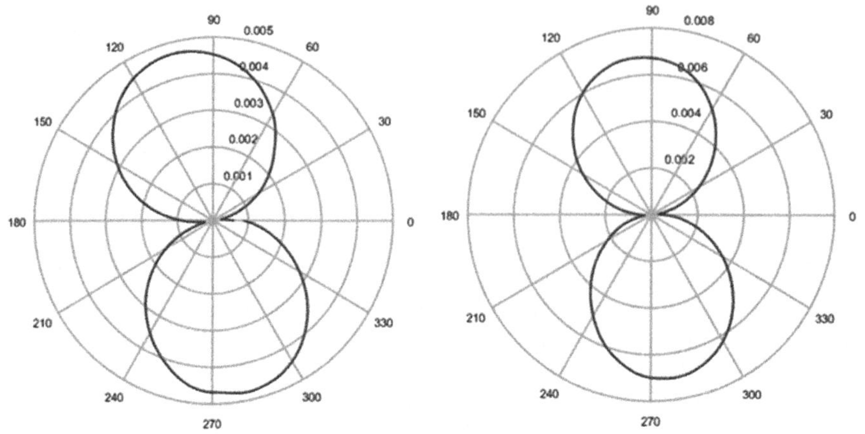

(a) **(b)**

Fig. 7.5 Measured radiant characteristics of the dipole

panels in these measurements. In summary, the dipole characteristic was measured by reflection from the conductive surface. The resulting characteristic is oval in shape, without the possibility of recognising the radiation characteristics of the measured antenna. This measurement was carried out to verify that the characteristics could not be measured by reflection from an ordinary conductive surface with comparable results.

In Fig. 7.5, the following is added: 96 is the characteristics of the dipole obtained—(a) by reflection through the panel; (b) by reference measurement. Depolarization panels can measure dipole antenna characteristics similarly to standard methods. However, the measurements showed the shape of the depolarization panel. Panel No. 2 provided the most accurate results as it took into account the shape of the characteristics and the level of the measured signal.

7.2 Optimization Layout Antennas with a Depolarizing Panel

To achieve the best possible measurement result, it was necessary to analyse the effect of the antenna position on the measurement result. Antennas used in damping chambers measure radiation characteristics in a remote field by positioning transmitting and receiving antennas opposite each other for direct visibility in both basic and standard distributions. If flat wave irradiation of the test antenna is required, reflection irradiates it from the parabolic reflector. This ensures the conversion of a spherical wave that is formed by a transmitter antenna to a flat wave. The reflector does not change the polarization of the wave, the receiving and transmitting antenna has the same polarization. This principle has gained wide acceptance.

Based on these two basic principles, 4 options for deploying antennas during measurement have been proposed. However, they all benefit from the reflection of wave from the depolarizing panel. In all cases, the antennas are located in one selected axis.

Layout:

1. The transmitting antenna is positioned in the rear axis of the receiving antenna, while the depolarization panel is exposed through the receiving antenna (as shown in Fig. 7.6a).
2. The transmitting antenna is located under the receiving antenna, the depolarization panel is irradiated from below (Fig. 7.6b).
3. The transmitting antenna is located under the receiving plate, separated from each other by a conductive surface (Fig. 7.6b).
4. The transmitting antenna is at the same height and plane as the test antenna, and the depolarization panel is irradiated from the side.
5. The transmitting antenna is located in the same position as in case 3, albeit with the antennas isolated from each other by a conductive surface (as depicted in Fig. 7.6c).

Antennas are separated by a conductive surface to ensure the reception of an inversely polarized wave. The phenomenon of cross-polarization is known to have a

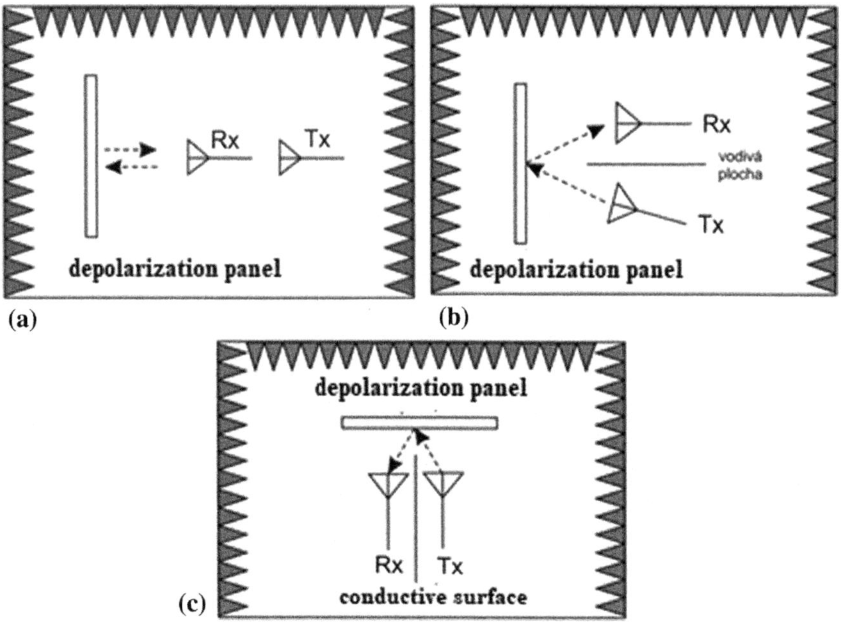

Fig. 7.6 Illustration of the antenna layout measuring

negative impact on the result. The antenna's design determines its ability to react to a wave with reverse polarization. As such, it may not manifest significantly with a symmetrical dipole. For directional properties of larger and more sensitive antennas, this effect can already be significant. These measurements used only one type of depolarization panel, specifically No. 8, with a symmetric dipole antenna. Using the comparative method, we selected the most suitable antenna layout from our results. The photo in Fig. 7.7 serves as an example of how we distributed the antennas.

Based on the measured characteristics, the following conclusions can be drawn. When measuring a symmetric dipole with a distribution of antennas (as shown in Fig. 7.7a), the characteristic is deformed (as seen in Fig. 7.8a). Comparison with the reference characteristic of the antenna clearly indicates a higher measured signal level, but the shape of the characteristic does not correspond to the Eight Shape pattern. Overflows in the minimum area are moderate, and the characteristic is not symmetrical around the 0–180° axis. Moreover, it does not have circular lobes.

A significant improvement was observed (as shown in Fig. 7.7b) by moving the transmitting antenna below the receiving antenna and adjusting the angle to the panel in such a way that the reflected signal is directed towards the receiving antenna.

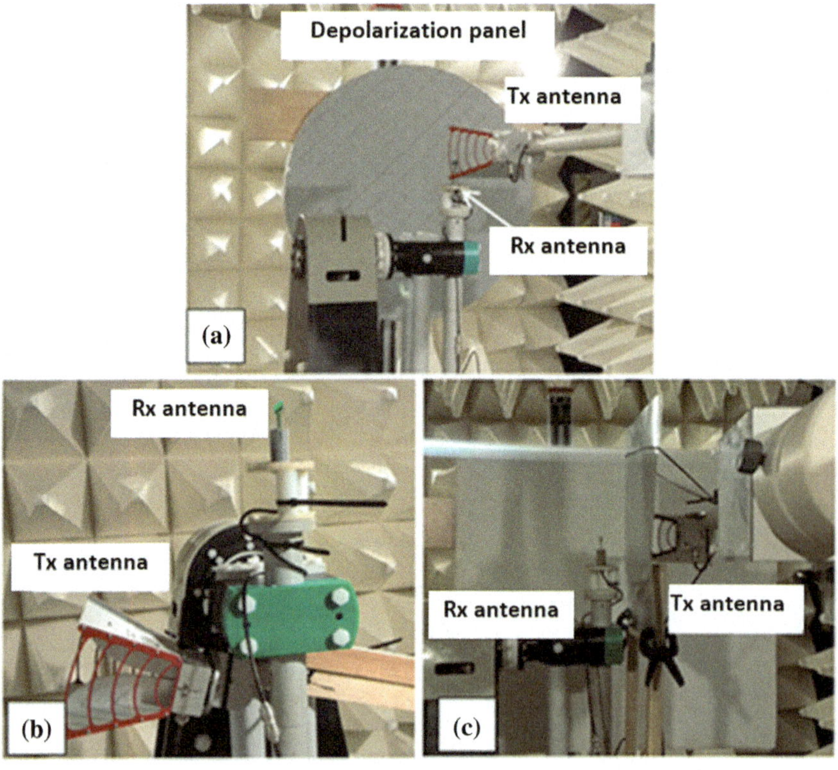

Fig. 7.7 Examples of different antenna layout options

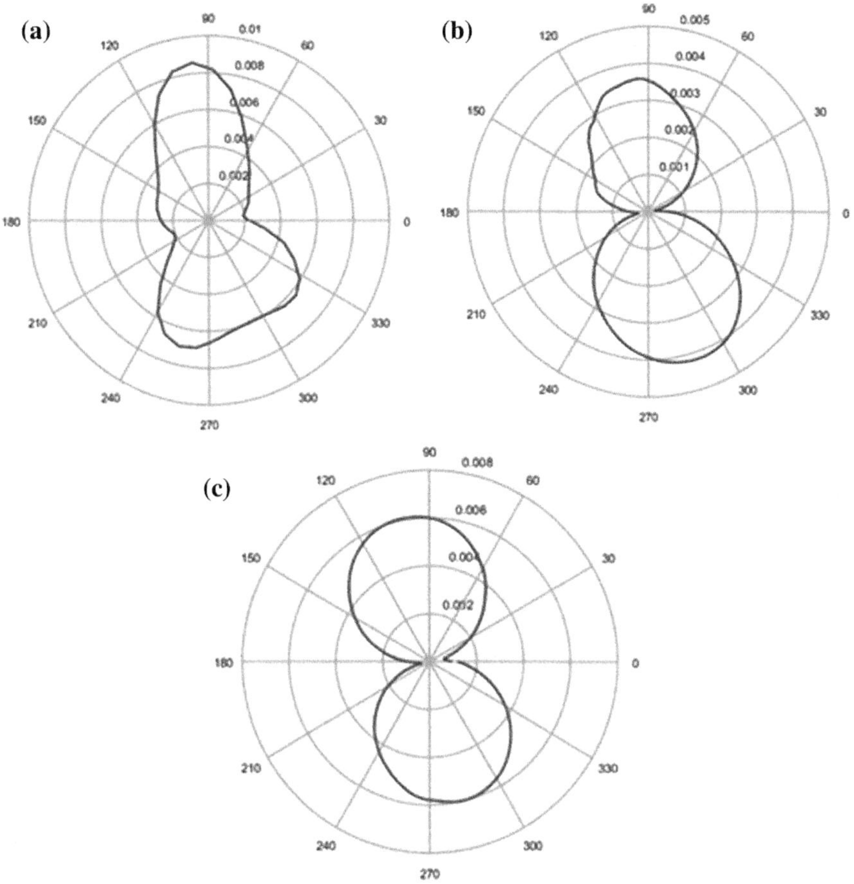

Fig. 7.8 Effect of antenna position on measurement, continuous results

The characteristic has already taken on the form of an eight, albeit distorted (see Fig. 7.8b). This implies that altering the antenna's position has a noteworthy impact on the measurement outcomes. The characteristic shows significant drops in the signal level in the region where antenna radiation is at a minimum. Nevertheless, the characteristic appears unbalanced. The lobes possess contrasting shapes and sizes. We assume there is an interaction between the antennas. As a result, we introduced a conductive surface between the antennas. Consequently, there was a decline in the measurement results. Compared to the measurements taken with the sheet, the signal level decreased. The characteristic does not allow for a definition of the areas with minimal radiation from the antenna, and it has an asymmetric shape. Each lobe has a unique shape and size. The hypothesis that the antennas interact with each other was not backed by evidence, and the proposed solution did not lead to any performance improvement.

To achieve such a distribution in antenna irradiation, the law of reflection of waves was maintained, and the transmitting antenna was positioned at an angle of approximately 20–25° to the horizontal surface. This steep angle causes uneven irradiation of the depolarization panel, which adversely affects its characteristics. Hence, placing the antennas one below the other was considered inappropriate and was not used again in any subsequent measurements. The latest antenna distribution was designed considering the uniformity of panel radiation (see Fig. 7.7c). The antennas were placed in one plane, which was also the plane of rotation of the receiving antenna, and positioned at the same height. This layout allows the transmitting antenna to be positioned almost perpendicular to the depolarization panel, ensuring nearly uniform irradiation of the entire area. The shape of the characteristic closely resembles the reference measurement. Although not prominent, the characteristic enables us to identify the antenna's minimum radiation areas. The upper lobe of the characteristic exhibits a slight deformation compared to the lower lobe. A minor level jump at 0° indicates a rotation in the manipulator arm of approximately 30° during the measurement of the dipole. The characteristic shape (Fig. 7.8c) and the measured levels indicate that the antenna's location is appropriate. To mitigate the interaction between antennas, a conductive sheet was placed between them. The shape of the characteristic is identical to the reference characteristic. The signal level is 0.001 mV lower than the reference measurement, resulting from the attenuation of the depolarization panel. The differences can be compensated by adjusting the transmitted power. As the output power of the generator was at its maximum level of 1 V, compensating for the difference was not possible. The lobes have a symmetrical, round shape. The minimum antenna radiation is indicated by sharp and clearly defined signal level drops to values near 0 mV. The characteristic is rotated about 5°, as in the previous case, due to the antenna's attachment in the manipulator's arm.

Based on the analysis of the results, it was concluded that the deployment of antenna number 3 is the most appropriate for the application. The results obtained from this antenna arrangement are the closest to the reference results in terms of the characteristics and signal level. All the subsequent antenna characteristic measurements were performed using antenna layout number 3.

7.3 Optimization Distance Antennas in Depolarization Panel

The analysis included examining the position of the depolarization panel relative to the antenna system and its impact on the radiation characteristics of the antenna being measured. Determining the optimal position was conducted in two stages. The first phase involved tilting and shifting the panel from the axis of the antenna array to evaluate its effect on the signal deflection relative to that axis. The second phase, which constituted the second experiment, was based on the assumption that the irradiation area of the antenna can change, creating unwanted phase differences in

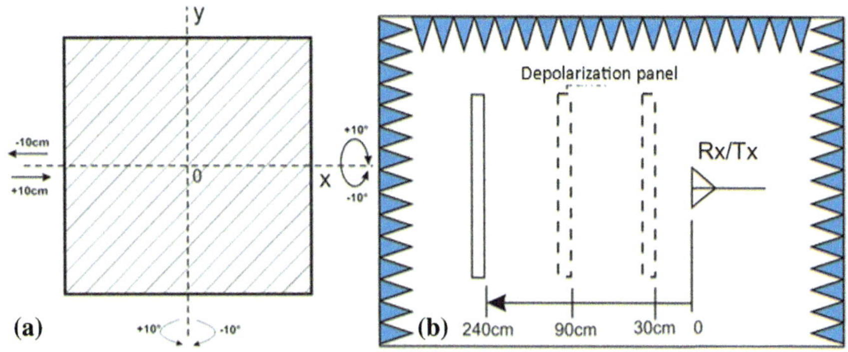

Fig. 7.9 Graphical representation measurement procedure

the signal, by varying the distance between the antenna and the depolarization panel. This phenomenon was described and confirmed in Sect. 6.5. Therefore, it is essential to determine the optimal panel distance from the antennas to minimize waveform deformation and ensure the characteristic shape is closest to the reference value. For easier geometric centre adjustment of the panel to the axis of the antenna array, type No. 8 was used again as a depolarization panel. The antenna under evaluation was a symmetrical dipole, and the antenna layout was No. 3, with the antennas arranged side by side and separated by a conductive surface. The reference characteristic used for comparison can be found in Fig. 7.5b. The antenna was positioned within the manipulator to ensure a minimum radiation characteristic of 90° at the given angle of rotation.

To change the panel's position relative to the antenna system's axis (as shown in Fig. 7.9a), the geometric centre of the depolarization panel was deflected through the panel movement. The depolarization panel shifted 10 cm to the left and then 10 cm to the right, tilted down by 10° and up by 10° in the horizontal plane, and finally turned 10° to the left and then to the right in the vertical plane. Significant differences were already visible with these deviations, making larger deviations unnecessary.

The panel's distance from the antennas (refer to Fig. 7.9b) was incrementally adjusted in 30 cm intervals within the 0–240 cm range. The 30 cm step size corresponded to the dimensions of the absorber within the attenuation chamber. The absorbers were gradually positioned on the chamber's floor between the antenna array and the panel. This method partially suppressed signal reflections from the chamber's floor and provided an accurate distance measurement, as the absorbers all had the same dimensions.

Figure 7.10 displays the selected measurement results. In the first course (a), the panel is rotated 10° to the left. In the course (b), the measurement is taken with the panel tilted 10° up, while in the last course (c), the panel is offset 10 cm to the left of the centre. Rotating the panel 10° to the left caused deformation of the characteristic outside the reference shape. The characteristic does not display clear minima. The characteristic is relatively symmetrical and has been rotated by 50°

from its initial position (where the characteristic had a minimum at 90°). The signal level was in proximity to the reference value. When the panel was rotated to the right by 10°, it resulted in a 30° rotation of the characteristic, relative to the base position. Nevertheless, the signal level is half of the reference value. The change in the panel's vertical plane rotation is believed to cause the reflected signal's main beam to deviate from the depolarization panel towards the rotated side. Reflection occurs in accordance with Snell's law of wave reflection. The antenna only captures a portion of the reflected signal.

When the panel is inclined at 10° in the horizontal plane, the effect is less pronounced since the panel is irradiated from the side. However, the angle of reflection relative to the receiving antenna remains the same, and the reflected signal is not deflected to the sides.

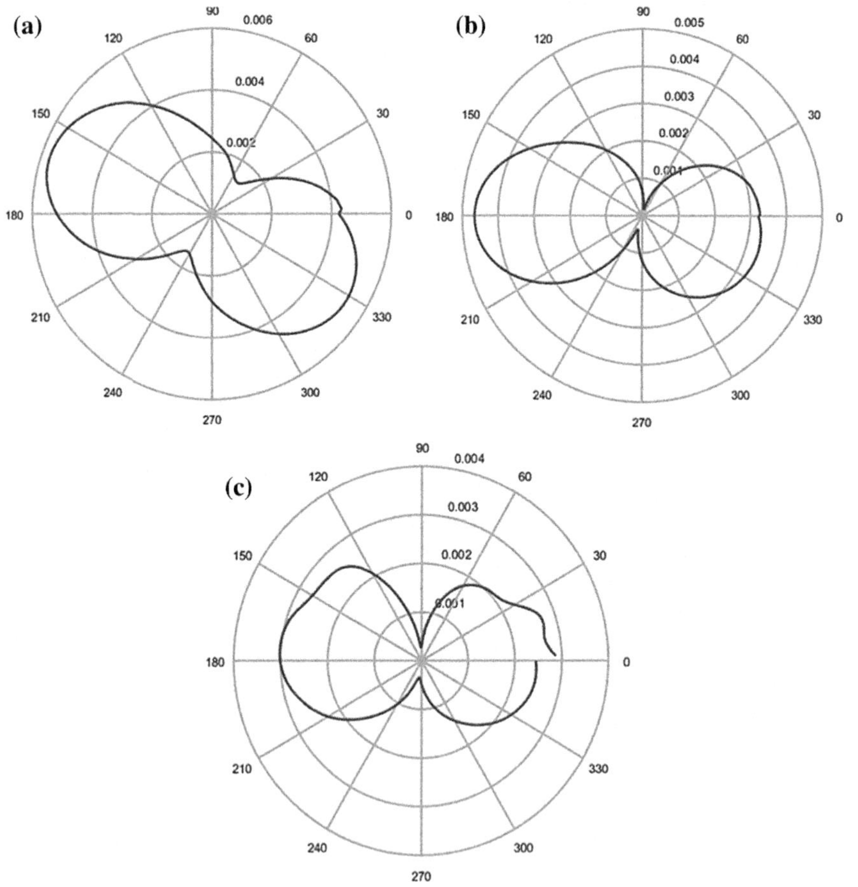

Fig. 7.10 Selected measurements when changing the position of the panel in the X/Y axis

The measured antenna is still irradiated by the main reflection from the panel, but its position in the vertical plane changes, its level decreases, and the resulting phase ratios are included in the measurement. The characteristics display lobes of asymmetric shapes and sizes. The subsidence in the minimum area is noticeable and clearly definable, a finding that was not as unambiguous when rotated in the vertical plane.

The most substantial impact on the resulting characteristics was observed when shifting the depolarization panel by 10 cm to the right or left. Shifting the panel leads to asymmetric irradiation of the conductor area. Although the law of reflection is observed, the incongruity of conductor radiation, caused by imaginary reduction or increase in the length of the conductors, results in a decrease in the depolarization phenomenon due to uneven phase ratios of the reflected wave.

The measured results show the necessity for ensuring a consistent geometric centre of the panel and antenna system along with zero rotation of the panel in the vertical plane.

The goal of the second experiment was to alter the panel's distance from the antenna system. Throughout the entire measurement, confirming the panel's correct position, as determined by the previous measurements, was essential. However, the handling of the system was challenging due to its large size and weight, and minor deviations were acceptable and, therefore, not considered as the effect of distance on the results.

The selected results of this measurement are shown in Fig. 7.11. The limit values are depicted in Fig. 7.9c. Measurement 1 was taken at a distance of 30 cm from the antennas of the panel, while measurements 2 and 3 were taken at distances of 90 cm and 240 cm, respectively. Figure 7.12 depicts all resulting characteristics. At a distance of either 90 cm or 120 cm from the antennas, the radiation characteristic experiences a slight deformation. The lobes exhibit the greatest symmetry in their characteristics and almost reach the reference value, with a difference of merely 0.001 mV. The characteristics allow for the unambiguous determination of both the minimum and maximum antenna radiation levels. Beyond a distance of 120 cm, not only does the signal level significantly decrease, but there are also noticeable distortions in shape. At a distance of 240 cm, the characteristics exhibit an undefined shape, and it becomes impossible to read the minimum and maximum radiation levels of the antenna.

It is believed that this phenomenon is due to the low level of the incident signal and the fact that the antenna irradiates the area outside the panel. This causes the depolarization phenomenon not to have such a level. The characteristics at 30 cm show similar characteristics to those at 90 and 120 cm, the signal level is low, which may be because the antenna only captures part of the main beam reflected from the panel. At such a distance, even a small change in the position of the antennas causes large changes in the reflected signal due to the small, radiated area of the panel and the large concentration of the signal radiating from the panel. These measurements indicate that the most appropriate distance between the antennas and the panel is in the range of 90–120 cm.

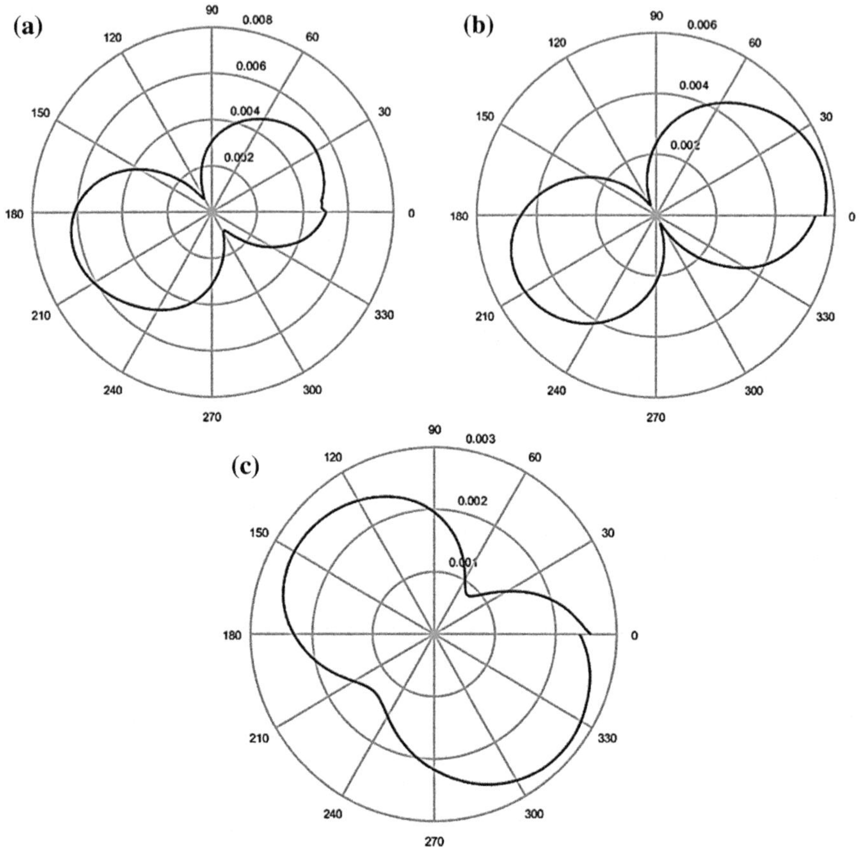

Fig. 7.11 Effect of antenna distance on the shape of the characteristic

7.4 Evaluation of Quality Directional Characteristics of Commercial Antenna

The characteristic of a symmetrical dipole provides an unambiguous view of the quality of the measurement. The shape of the characteristic of a symmetrical dipole is unequivocal and has defined minima and maxima of radiation. A detailed analysis of the results is not necessary, since the success of the measurement can be seen at first glance. In the previous section, the possibility of using a depolarization panel to measure frequency characteristics was confirmed.

Three SANO reflector antennas [1, 2] were selected to verify the functionality of the panel for other antennas, such as the symmetrical dipole. Before measuring the radiation characteristics, VSWR waveforms were calculated for all three antennas, and then reference radiation patterns were measured using the standard antenna

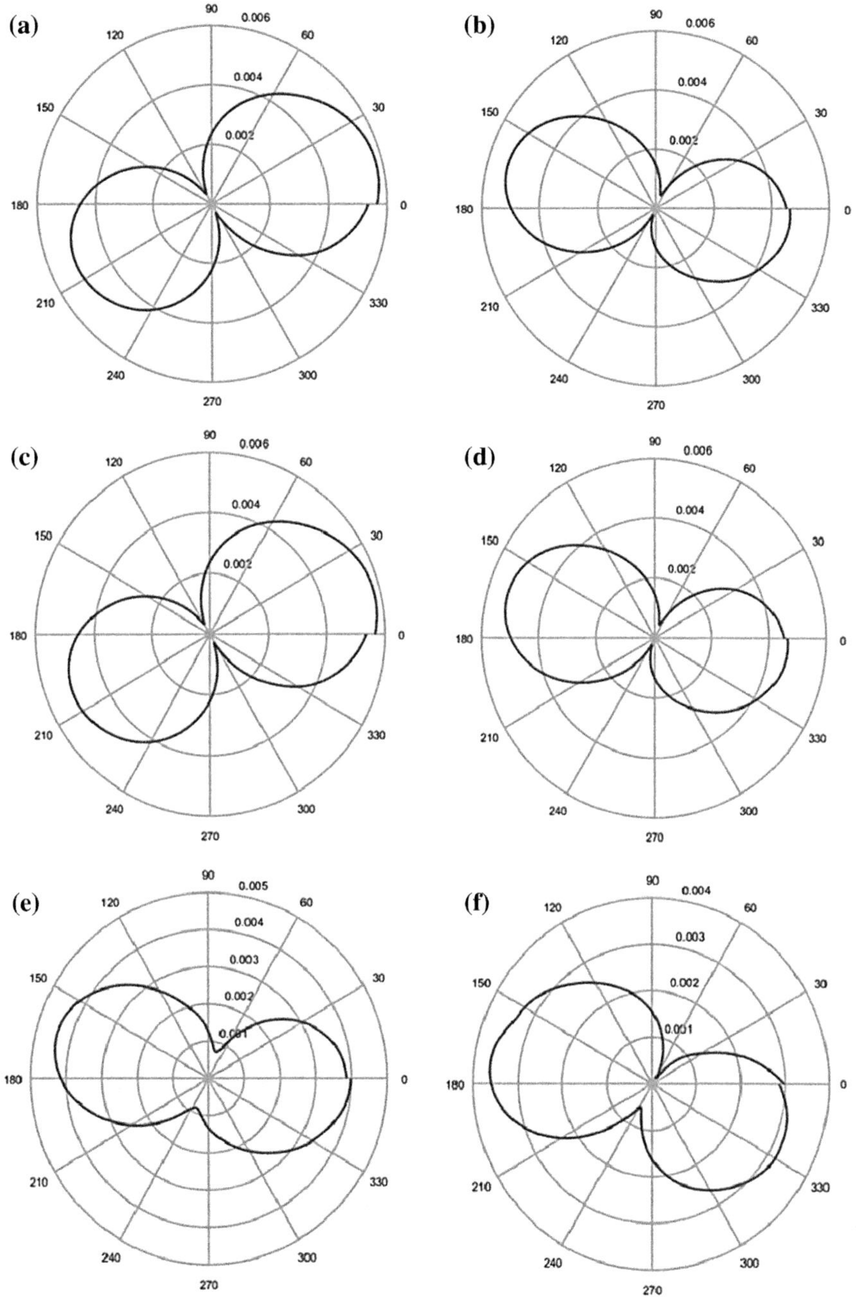

Fig. 7.12 Effect of panel distance on measurement

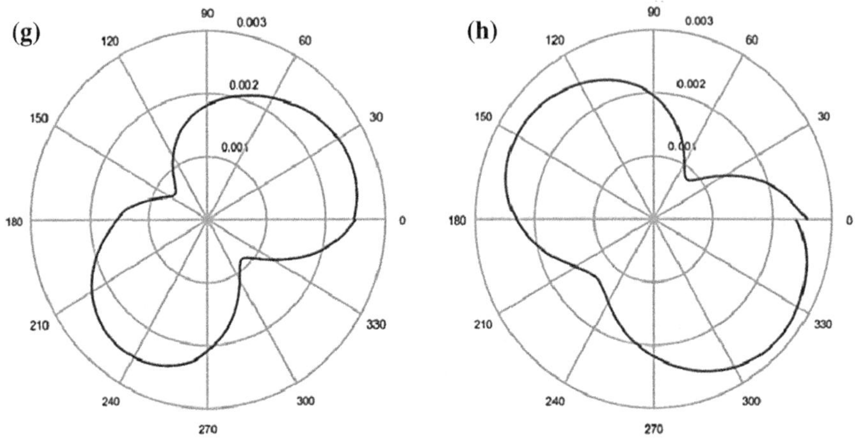

Fig. 7.12 (continued)

measurement method. Reference values of the radiation diagrams in Annex D, Figs. 9–11.

Dimensionally, the smallest of them is labelled SANO MICRO 8 (Fig. 7.13(1)). According to the information available in the catalogue, this antenna has a working range of 2.4–2.475 GHz [4]. The 2.2–2.57 GHz range was established from the VSWR test, taking into account frequencies with VSWR < 2. Analogy evaluated with other antennas. The antenna from Fig. 7.13(2) has an unknown designation. Compared to MICRO 8, it has a larger diameter and is composed of four illuminators with one reflector placed in a square. As catalogue data could not be found for this antenna, so the measured VSWR could not be compared with them. The VSWR course is at No. 106th Meeting of the Conference of the Parties to the Basel Convention.

This measurement defines the working area of the antenna from 1.7 to 2.7 GHz. The last measured antenna (Fig. 7.13(3)) bears the designation SANO SECTOR V16. It is an antenna of elongated shape with four illuminators on one reflector. Opposite the passing round antenna, it has illuminators in a row next to each other. The working area defined in the catalogue is between 2.4 and 2.47 GHz. The measured work area ranged from 2.1 to 2.6 GHz. For all three antennas, the required measured frequency is 2.4 GHz in the operating range of the antennas (Figs. 7.14, 7.15 and 7.16).

After measuring the VSWR, the radiation characteristics of these antennas were measured. The catalogue data showing the directional pattern of the antenna is only available for the SANO MICRO 8 antenna. The radiation pattern of the other two antennas is unknown.

The radiation pattern of the SANO MICRO antenna is measured using a standardized measurement method (reference value) (Fig. 7.17a). The measured diagram (Fig. 7.17a) is used as a reference for further measurements. Antenna SANO SECTOR V16 and the unknown antenna do not have catalogue radiation patterns.

Fig. 7.13 Reflector antennas used as measuring objects

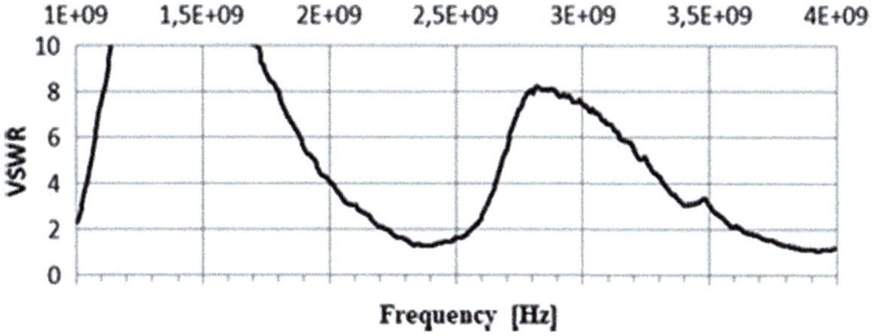

Fig. 7.14 VSWR SANO MICRO 8 courses

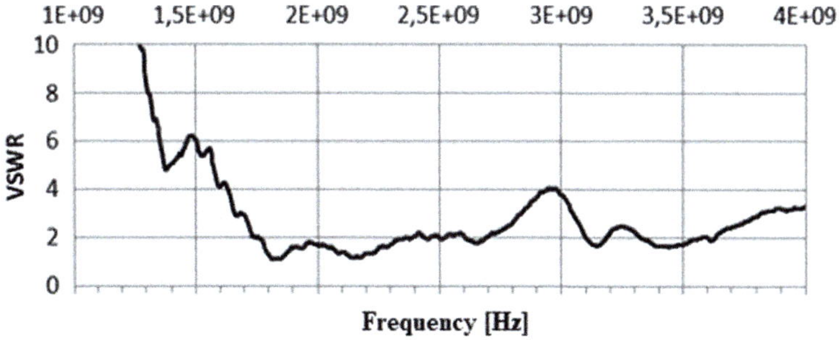

Fig. 7.15 VSWR course of a large round antenna

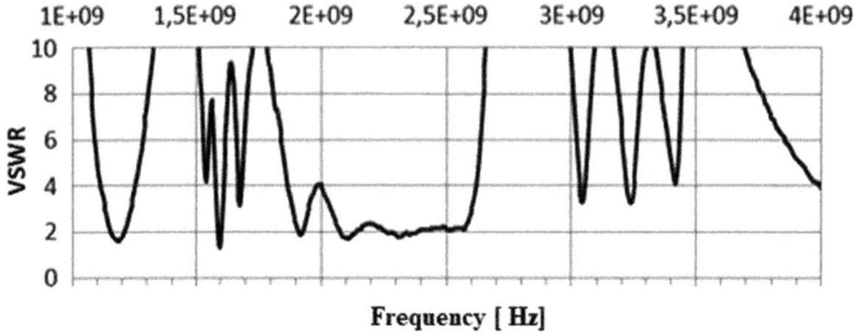

Fig. 7.16 VSWR SANO SECTOR V16

The characteristics were obtained by standard measurement. The SANO SECTOR V16 antenna has a radiation pattern (Fig. 7.17b). This pattern is comparable to that of the SANO MICRO 8. The main radiation lobe is more concentrated than that of the SANO MICRO 8, and the field level reaches a maximum of 0.014 mV. In addition to the main radiation lobe, the antenna has two lateral lobes at angles of 150 and 45°, which are asymmetrical to each other. A large circular antenna with four radiators has the narrowest main radiation lobe of all antennas (Fig. 7.17c). The antenna is clearly directional, with a very small beam angle. The measured maximum field level of the main radiation lobe was 0.018 mV. From the analysis of the radiation characteristics described above, the SANO MICRO 8 antenna was selected as the object of measurement using the depolarization panel. Its characteristic has a convenient shape for control measurements.

Measurements of the symmetrical dipole using a depolarization panel have shown that the results obtained using different depolarization panels are comparable. The fundamental differences were in the maximum level of the main lobe and the asymmetric shape of the characteristic. The results obtained by measuring with a depolarization panel are shown in Fig. 7.18, where (a) with Panel No. 5, (b) with Panel No. 8, (c) using Panel No. 2, and (d) using Panel No. 1. The radiation characteristics obtained by depolarization of Panels No. 5, No. 8 and No. 2 have the same character. The shape of the measured curve is identical to the reference curve (Fig. 7.17a). The field level in the main radiation lobe is comparable in all cases and ranges from 0.013 to 0.014 mV. Slight differences are due to the attenuation of the panel. Radiation lobes in the 180–360° half-plane is ignored because the antenna is rotated by the back of its reflector to the depolarization panel.

The characteristic curve obtained by depolarization Panel No. 1 shows the greatest deformation of the shape. The radiation pattern is deformed towards 120°, where the signal level increases, and on the opposite side, it decreases. This panel also shows some distortion of the radiation pattern when measured on a symmetrical dipole. Measurements of the directional radiation pattern (Fig. 7.18) only confirm the fact that the depolarization panel can be used both for non-directional patterns of the symmetrical dipole type and for directional patterns with a reflector. The results obtained by reflection of an electromagnetic wave from a depolarization panel are

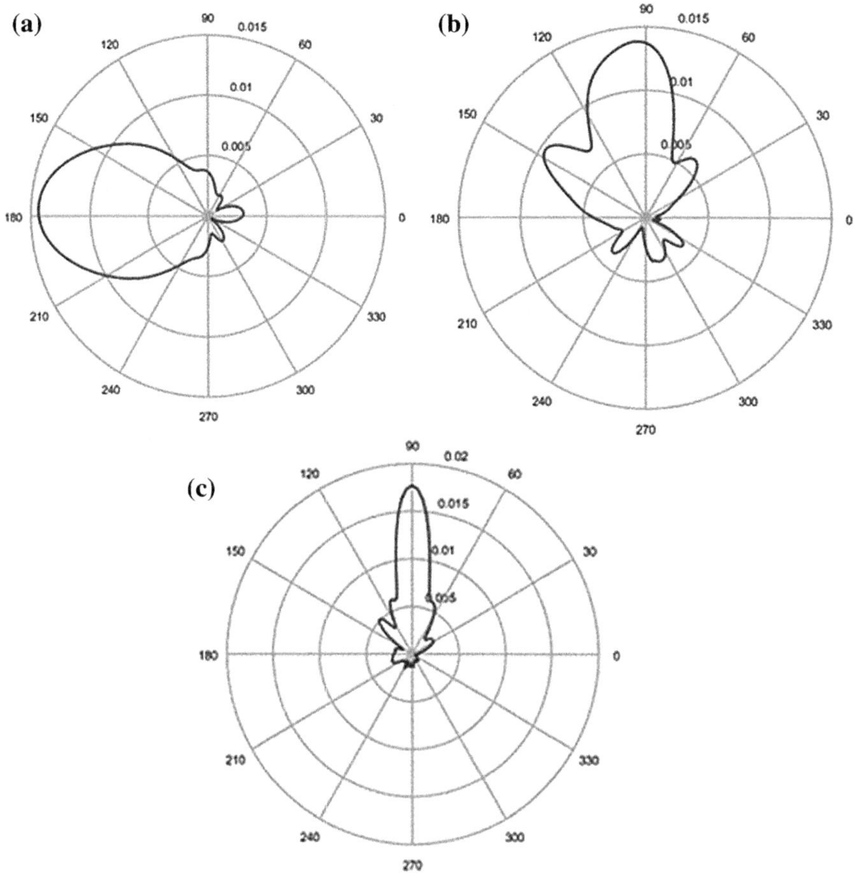

Fig. 7.17 References measurements of directional antenna. **a** SANO MICRO 8; **b** SANO SECTOR V16; **c** unknown with 4 sources

comparable in shape and level to the result obtained by a reference measurement in a standardized manner.

7.5 Directional Characteristics of Dipoles Measured in the Laboratory

All previous measurements were performed in the anechoic chamber. In such measurements, the antenna system and the entire measurement process are not affected by external environmental influences such as interference from surrounding sources and reflections from surrounding objects. The depolarization panel is

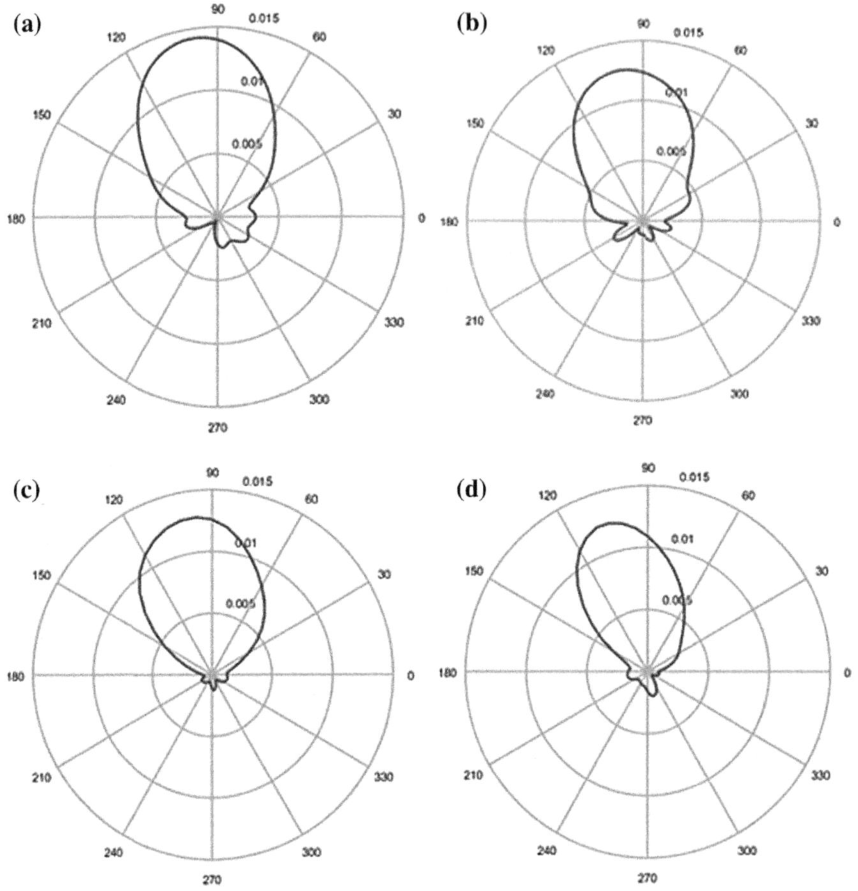

Fig. 7.18 Characteristics of SANO MICRO 8 obtained by reflection from the panel

designed to allow measurement of the radiation characteristic even in such an imperfect environment. This property has been experimentally verified. Laboratory equipment from the attenuation chamber was moved to the laboratory room where the experiment was conducted. The measurement was affected by external influences. On the laboratory premises, there are Wi-Fi sources, cell phones and radio receivers. These sources of interference were not shielded or turned off during the measurements. A major source of interference was the anechoic chamber itself since the measurement was made from the outside. The walls of the anechoic chamber are made of galvanized sheet metal on the outside and form a massive reflective surface of about 7×3 m. There were metal tables and chairs in the laboratory, and on the opposite wall from the wall of the anechoic chamber, there were fixed metal writing boards. The laboratory was considered sufficiently disturbed for the radiation pattern measurement experiment. The measured antenna was a symmetrical dipole for the reasons described above. The directional antenna was not measured. The distribution of the elements during the measurement is identical to that in the attenuation chamber. The panel was placed at a distance of 1 m from the antenna system. The transmit and receive antennas were in one plane and separated by a conductive surface. Selected results are shown in Fig. 7.19. As the first measurement, the control measurement was performed in a standardized manner (Fig. 7.19a). The characteristic has an irregular shape; it is not symmetrical, and it is impossible to determine the minimum of antenna radiation from it. On the characteristic, distortions appear in the area where signal minima are expected. The shape cannot be considered an octagon, as is the case with the reference characteristic (Fig. 7.19a). This measurement confirms the assumption that there is a sufficient amount of external interference and potential sources of wave reflection in the laboratory.

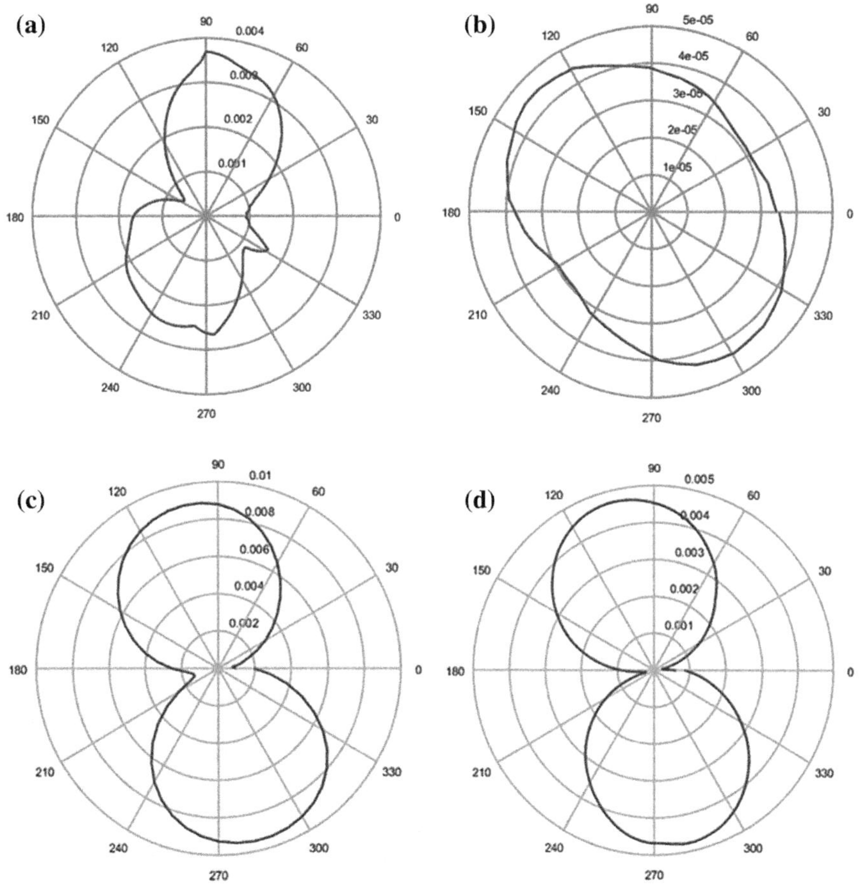

Fig. 7.19 Directional characteristics of the dipole outside the anechoic chamber

After this control measurement, the elements were placed as described above. For the measurements, Panels No. 1, No. 2, No. 5, and No. 8 were used.

Then, after measurement by the standard method, the radiation characteristic was measured by reflection from the conductive surface/sheet (Fig. 7.19b). The characteristic has an oval shape, and it is impossible to describe the shape of the curve more precisely or to find a similarity with the reference characteristic. Thus, it was found that it is impossible to measure the octal characteristic by reflection from the conductive surface under the given conditions. Measurement Fig. 7.19c is the measurement by depolarization Panel No. 2.

The curve in Fig. 7.19d represents the measurement obtained by reflection from Plate No. 2 in the attenuation chamber. Comparing these two traces, the differences are not very obvious. A significant difference is in the minimum signal value in the area where the antenna has minimum radiation. In the case of measurements in the attenuation chamber, values close to zero are obtained in this area, whereas in the case of measurements outside the chamber, the values in this area are not as sharp and pronounced. Nevertheless, the characteristics are clearly definable and can be considered correct, since the signal level and the unambiguous shape of the waveform are comparable to the reference value.

The shape of the measured dipole characteristic is comparable for all panels (Fig. 7.20), and there is no significant deformation similar to that seen in standard method measurements. From the characteristics, it is possible to determine the maximum and minimum radiation of a symmetrical dipole. The measured characteristics should have a clearly legible and defined figure-of-eight shape. Dips in the area of the minimum radiation are sharp, and at Panel No. 1 are close to zero. In the case of Panel No. 8 (Fig. 7.20c), there is again a significant decrease in the level to below 0.0015 mV. However, the shape of the curve remained undistorted. A closer analysis of the characteristics shows a slight deformation of the lower lobe in the case of the measurement with Panel No. 5 (Fig. 7.20b). The different size of the lobes in this panel was also reflected in the attenuation chamber measurements. The results obtained with the depolarization Panel #2 (Fig. 7.20d) are identical to the reference measurements, as in the previous measurements, and do not show any distortions or significant level drops caused by the attenuation of the panel, as was the case with Panel #8. The characteristics (Fig. 7.20e) were obtained by reflection from depolarization Panel No. 1. As in the previous cases, the characteristic is not significantly deformed. Measurements outside the anechoic chamber, in an environment with various sources of interference, should demonstrate the resistance of this method to external influences.

The measured characteristics are identical to those measured in the attenuation chamber and there is no deformation caused by external interference from various sources. Therefore, it can be assumed that the characteristics of the depolarization panels are not affected by external sources of interference. This assumption is valid only under laboratory conditions. Interference sources may cover a wide range of frequencies and power levels, and it is not possible to verify the properties of depolarization panels for such a wide range.

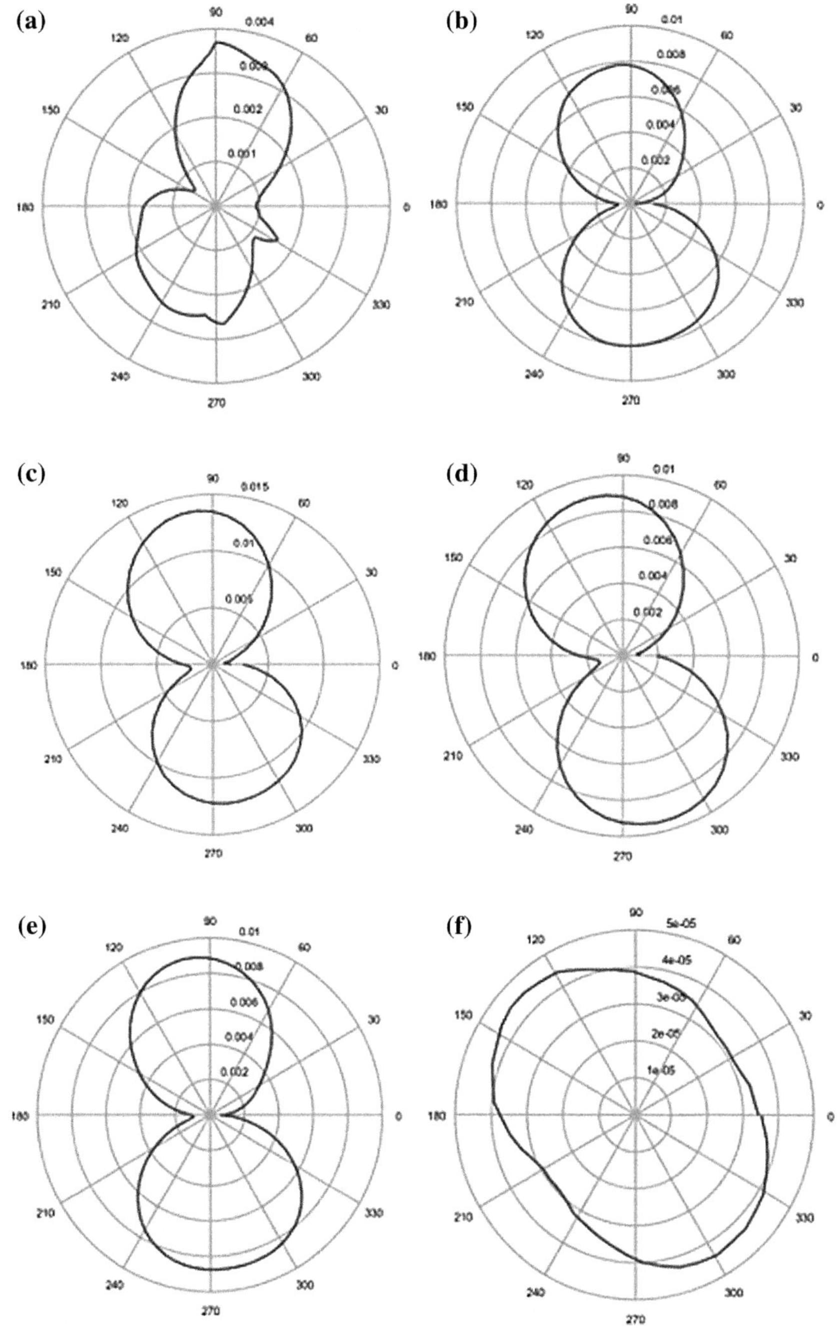

Fig. 7.20 Measured radiant characteristics of the symmetrical dipole

References

1. Chung, B.K., Chuah, H.T.: Modelling of RF absorber for application in the design of anechoic chamber. Prog. Electromagn. Res. PIER **43**, 273–285 (2003)
2. https://www.sanoantennas.com/sk/SANO-Micro-8
3. Garg, R.: Analytical and Computational Methods in Electromagnetic. (2008). ISBN 9781596933859
4. Labun, J., Šaňo, I.: Anténa pre počítačovú komunikáciu, patentový spis č. SK 4439, 4 s. ÚPV SR, Banská Bystrica (2006)

Chapter 8
Exploitation Methods in Antenna Technology

After the implementation of the depolarization panel, the optimization of the attenuation depolarization chamber and the design of a new method for measuring the radiation characteristics of antennas, the actual measurement of the antenna characteristics was carried out.

For the purpose of measurement and objective evaluation of the measurement results, an antenna was selected, the characteristics of which are generally known. It is an antenna of the half-wave symmetric dipole type, whose radiation characteristic in polar coordinates has the shape of two circles arranged in a mirror image with respect to the longitudinal axis of the dipole, in the shape of a symmetric eight. Such a radiation pattern is considered a standard for the evaluation of qualitative measurement results (Fig. 8.1). The measured half-wave dipole was designed at a frequency of 2.4 GHz and realized on a photosensitive double-sided printed circuit board with a copper thickness of 35 μm, the material of the dielectric is FR4 laminate, also suitable for RF applications.

8.1 Qualitative Evaluation Experimental Measurements

The first measurement of a symmetrical half-wave dipole was carried out in a classical anechoic chamber in the standard way, i.e., direct irradiation of the measured antenna with a reference transmitting antenna. The purpose of this measurement was only to confirm whether the classical method of measurement achieves the desired result, i.e., the characteristics of the symmetrical figure-of-eight shape. The result of measurements in the attenuation chamber using the classical method is a symmetrical figure of eight, which corresponds to the theoretical assumption (Fig. 8.2).

A second measurement of the symmetric half-wave dipole was also performed in the attenuation chamber. A special depolarization attenuation chamber was created in

M. Krchňák et al., *Advancements in Antenna Measurement*, Lecture Notes in Electrical Engineering 1108, https://doi.org/10.1007/978-3-031-48835-1_8

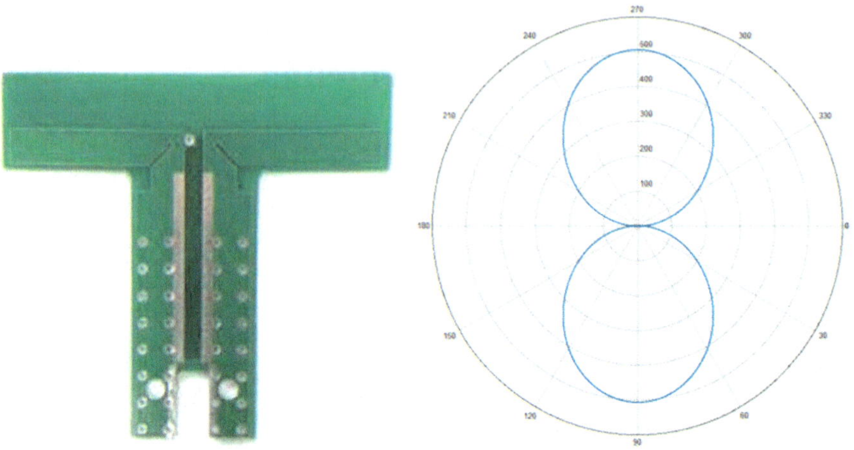

Fig. 8.1 Measured the dipole and its calculated radiant characteristics

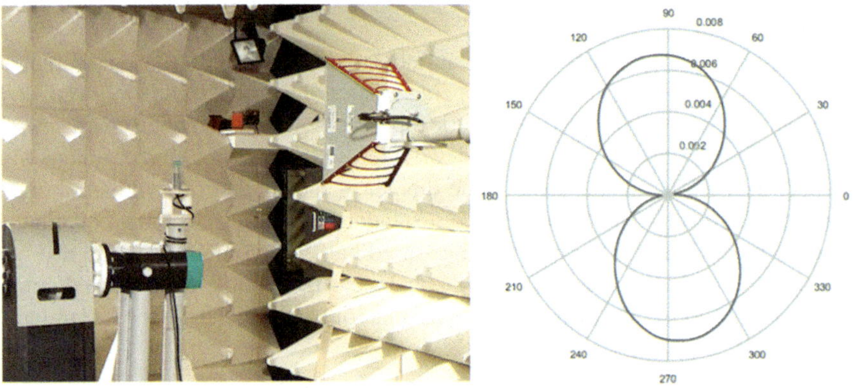

Fig. 8.2 Measurement in the anechoic chamber by the classical method

its premises. The measurement itself was carried out by an indirect method of irradiation of the measured receiving antenna by a laboratory funnel transmitting antenna through a depolarization panel. The purpose of this measurement was to confirm whether the new measurement method using the depolarization panel would give the same results. In this case, a depolarization attenuation chamber was used, but it was located in an ideal attenuation environment without interference. This is a confirmation of the function of the new method itself, whether the resulting characteristic will be the shape of a symmetrical figure of eight. The result of measurement in the depolarization chamber located in the attenuation chamber, when measured by the indirect method of irradiation of the measured receiving antenna by the transmitting antenna through the depolarization panel, is shown in Fig. 8.3.

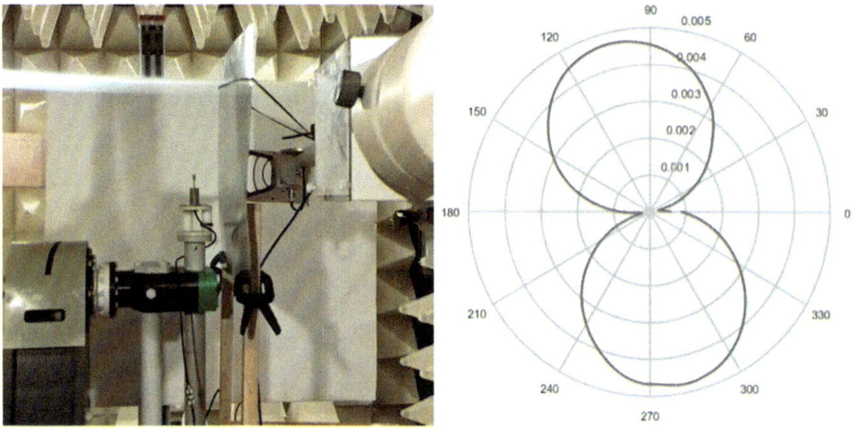

Fig. 8.3 Measurement in the chamber by the depolarization method

In this case, too, a symmetrical figure of eight was measured, which corresponds to the theoretical assumption.

The third and further measurements of the symmetrical half-wave dipole were carried out in the classical laboratory, i.e. outside the interior of the attenuation chamber. The measurement was carried out in the laboratory in a classical way, i.e. by direct irradiation of the measured receiving antenna by a laboratory funnel transmitting antenna. The purpose of the measurement was to show how the shape of a symmetrical figure-eight is distorted when measured by the classical method indoors in the presence of different conductive reflecting surfaces. The result of the classical laboratory measurement is a significantly deformed asymmetric figure-eight whose shape corresponds to the distribution of parasitic reflective conductive surfaces in the laboratory itself (Fig. 8.4). From the point of view of the measured characteristic, its deformed shape corresponds to theoretical assumptions, since when an antenna is measured in any environment, some deformed characteristic is always measured. However, the characteristic measured in this way does not correspond to the radiation of the antenna, but to the distribution of the conductive parts, their shape and the electrical properties of the internal equipment of the laboratory. In addition, the measurement is also influenced by the walls of the laboratory and the location of the measuring station in the limited space of this laboratory.

The fourth measurement of the symmetrical half-wave dipole was also performed in the laboratory. The measurement itself was performed by an indirect method of radiating the measured receiving antenna. The reflection was made by an ordinary sheet metal plate. The measurement was intended to show the result obtained by an indirect method of measurement using an ordinary conductive reflecting plate. The result of the measurement in a classical laboratory, outside the attenuation chamber, obtained by the indirect method, is again a significantly deformed asymmetric eight, the shape of which corresponds to the distribution of parasitic reflective conductive surfaces in the interior of the laboratory (Fig. 8.5). In this measurement, however, the

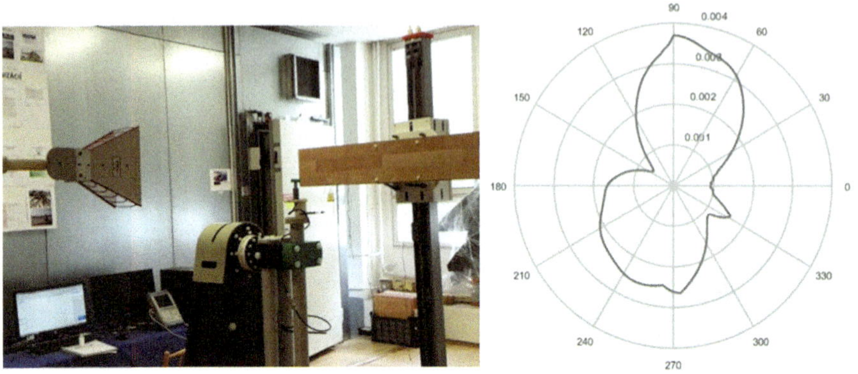

Fig. 8.4 Measurement in the laboratory by direct method

result was more negatively influenced by the nearby conductive reflective panel itself, in front of which the measured antenna rotated during the measurement. Therefore, the radiation pattern in this case has a deformed, almost circular shape.

The two measurements (third and fourth) carried out in a classical laboratory showed that such an environment is not suitable for measuring radiation patterns with a standardized method. For this reason, a depolarization attenuation chamber was created.

The fifth measurement of the symmetrical half-wave dipole was also carried out in the classical laboratory. The measurement itself took place in the laboratory, but already a new, so-called, by indirect method of irradiation of the measured antenna through a depolarization panel. The main purpose of this measurement was to confirm or disprove a new theory of possible suppression of parasitic reflections in the measurement of directional characteristics of antennas using the attenuation of cross-polarization between the measured and the measuring antenna. The result of

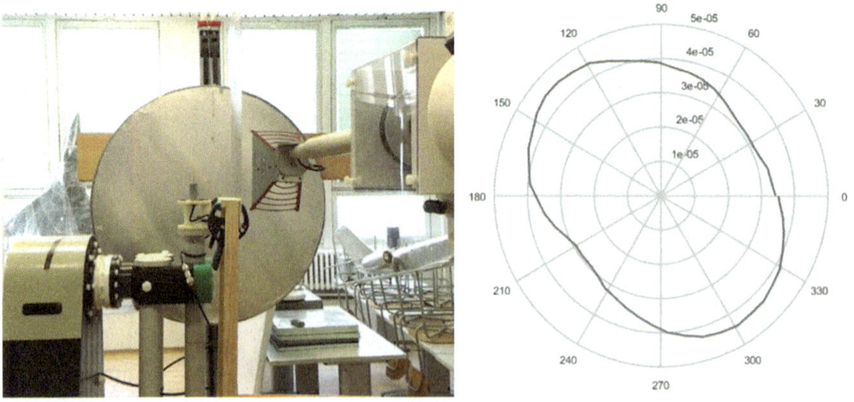

Fig. 8.5 Measurements in the laboratory by an indirect method with plate metal

the measurement in a classical laboratory, outside the attenuation chamber, measured by an indirect method with a depolarization panel, is a quality symmetrical figure of eight, whose shape corresponds to the measured values inside the classical attenuation chamber (Fig. 8.6).

The result of the measurement confirmed that the measurement of the directional characteristics of the antennas by the new method presented can also be carried out in an ordinary laboratory, office or other space where the interference is located.

The depolarizing method presented for measuring the characteristics of antennas can eliminate these parasitic reflections to such an extent that the results thus measured correspond to the measured results of the classical anechoic chamber. Progress Fig. 8.7a displays measurements in the classical anechoic chamber and Fig. 8.7b, the result obtained by measurement on the premises of the laboratory.

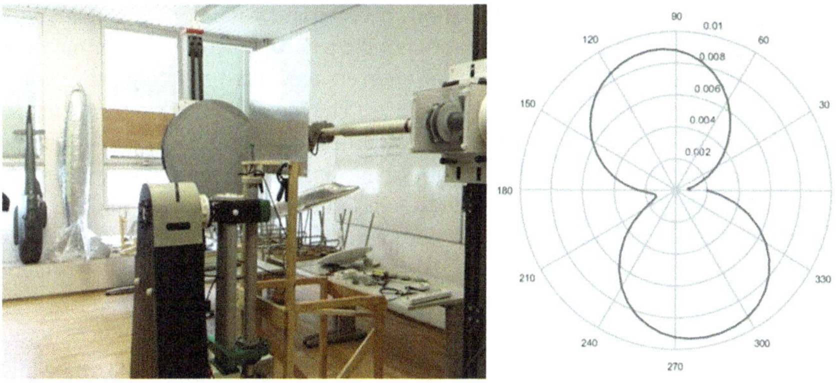

Fig. 8.6 Measurement in the laboratory by depolarization panel method

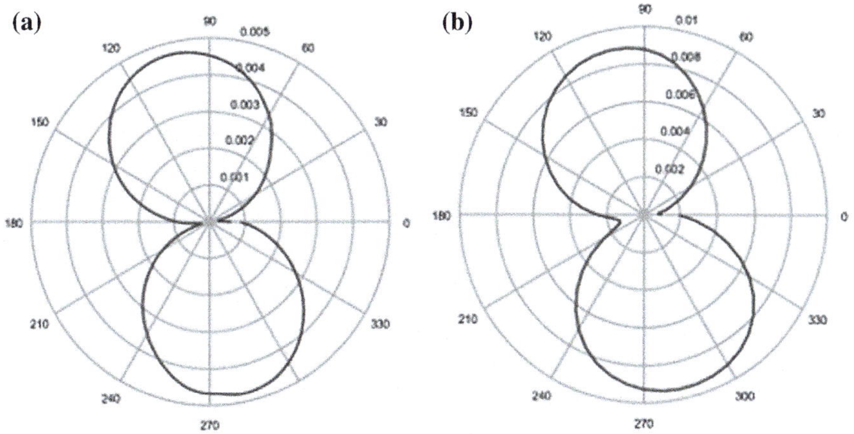

Fig. 8.7 Measurement results in anechoic chamber and laboratory

8.2 Recommendation for the Design and Use of the Depolarization Chamber

The preceding and subsequent chapters covered extensive and time-consuming measurements utilising a depolarisation panel. These measurements were imperative in determining the properties of the depolarisation panel. The measurement method and methodology were created explicitly for use with the depolarisation panel. The work involved the gradual development of experiments, measurements, and their associated procedures. The objective of all experiments was to procure measurements of the radiation characteristic as close as possible to the reference measurement in terms of qualitative attributes.

Today, attenuation chambers, which are expensive and require more space, are used to measure the radiation features of antennas. The depolarisation panel was created specifically for testing the radiation characteristics of antennas under laboratory conditions. Laboratory conditions are not the perfect environment for measurements because they are affected by external interference from multiple sources, which can include waves reflected off walls or objects. The measurement system equipped with a depolarisation panel can tolerate environmental interference to some degree. The cost and complexity of the depolarisation panel is minimal in comparison with the attenuation chamber. The results from both standardized measurement and measurement using a depolarization panel show comparable qualitative outcomes. Based on these substantiated facts, it can be argued that the depolarization panel will facilitate the measurement of simple directional characteristics in the laboratory without requiring an attenuation chamber or a special "non-reflective" laboratory modification. From the results of the measurements, it is possible to make recommendations for the methodical measurement procedure and the design of the depolarization panel/chamber. Following the recommendation in this section, it ought to be possible to implement a depolarization panel for the required operating frequency and conduct measurements for the directional characteristics of less complex antenna systems. Antenna systems with high complexity, like phase field or other directional antennas, commonly exhibit characteristics that entail complexity. It is even difficult to measure such a characteristic in an attenuation chamber. As complex systems have not been measured, only simple directional antennas and symmetrical dipoles are recommended to use as a depolarization panel.

8.2.1 Recommendation and Technical Arrangement of Depolarization Chamber

Chapter 5's experimental results yield the following recommendations for designing the panel. The depolarization panel comprises a dielectric, a conductive surface, and a reflective conductive grid. The panel's design affects its frequency and attenuation properties, impacting the characteristics' measurement quality. The technological possibilities to manufacture such a panel are currently boundless. All depolarization panels made to date have been created using cost-effective materials and are connected through gluing. A fundamental and primary parameter that initiates the entire production process is the operating frequency to which the depolarization panel is calibrated. The frequency corresponds to the dielectric thickness, which can be ascertained by referring to Eq. (6.5) in Sect. 6.4. The dielectric material influences the attenuation properties of the panel and the extent of the depolarization phenomenon. The geometric dimensions and shape of the panel also impact the measurement results. Two primary panels were produced—a circular and a square panel. The measurements illustrated the sharp edges of the panels, which were less noticeable when circular in shape. The appropriate size of the panel must be chosen based on the radiation characteristics of the transmitting antenna. Irradiating the panel with a plane wave would be the most suitable method. The size of the panel is not considered essential in this case. Depolarization panels with other geometric shapes have not been manufactured or tested. An aluminium sheet was used as the conductive surface of all depolarization panels. Since the electromagnetic wave gets reflected from the panel surface, the conductive material's properties are not deemed crucial. A conductive grid is present on the panel surface. The latter comprised a system of parallel wires with uniform thickness and fixed distance between the conductors. Chapter 5 presented measurements dedicated to the arrangement of conductors and the gaps between them. Measurements have demonstrated that the gap between the wires has to be chosen based on the operational frequency requirement of the panel. For frequencies below 2 GHz, it is advisable to opt for a wider gap between the wires. For higher frequencies, a narrower gap is preferable. The impact of conductor material or insulation on the outcome was challenging and, hence, not investigated. The results of the experiments in Chap. 5, the following recommendations are made for the design of the panel. The depolarization panel is a composite material consisting of three basic parts—a dielectric, a conductive surface, and a reflective conductive grid. The design of the panel affects its frequency and attenuation properties, which in turn affects the quality of the measurement results. The technological possibilities of manufacturing such a panel are currently unlimited. All the depolarization panels produced so far were made of cheap materials and connected by gluing. The first and basic parameter, on which the whole production is based, is the working frequency, to which the depolarization panel is tuned. The frequency is based on the thickness of the dielectric, which can be determined by reference Eq. (6.5) in Sect. 6.4. The dielectric material influences the attenuation properties of the panel and affects the size of the depolarization phenomenon. The geometric size and shape of the panel also affect the measurement results. Two basic panels were made—a round panel and a square panel. The measurements showed sharp edges of the panels, which were not so pronounced with a round-shaped panel. The size of the panel must be selected according to the radiation characteristics of the transmitting antenna. The most appropriate would be to radiate the panel with a plane wave. In this case, the size of the panel is not so important. Other geometric shapes of depolarizing panels have not been manufactured or tested. In all cases, the conductive surface of the depolarizing panel was in the form of an aluminium sheet. The electromagnetic wave is reflected from the surface of the panel, so the properties of the conductive material are not critical. On the surface of the panel, there is a conductive grid. The grid is formed by a system of parallel wires with a constant thickness and distance between the conductors. The measurements in Chap. 5 were devoted to the arrangement of the conductors and the gaps between them. The measurements proved that the spacing between the wires must be chosen according to the required operating frequency of the panel. For low frequencies below 2 GHz, it is recommended to choose a larger distance between the wires. For higher frequencies, a smaller gap is preferable. The effect of the conductor material or the effect of the conductor insulation on the result has not been investigated due to its difficulty.

8.2.2 Recommendation of Measurement Methodology in Depolarization Chamber

Along with the development of the design of the depolarization panel, the measurement methodology using this panel was also developed. The measurement methodology used for measurements in attenuation chambers could not be used. Measurement with a depolarization panel requires irradiation of the panel with a transmitting antenna. The antenna to be tested is not irradiated directly by the transmitting antenna, as is the case with the standard measurement, but is irradiated by the reflected wave from the depolarization panel. This wave has the same polarization as the antenna being tested. The wave leaving the panel has an inverted polarization of 90° relative to the wave incident (from the transmitting antenna). Thus, the transmitting antenna must have a polarization perpendicular to the tested polarization of the tested antenna. The measurement methodology has undergone a process of development and the measurements have resulted in some basic advice for the placement of the laboratory and components during the measurement.

Measurement of the directional characteristic takes place by irradiating the test antenna with a wave that is reflected from the depolarization panel. As in standardized measurements, a measuring and measuring antenna must be present. The frequency range of the transmitting antenna must overlap with the antenna being tested. The depolarization panel must be tuned to this frequency. The requirements for laboratory equipment required for these measurements do not differ from those for standardized measurements. In both cases, a signal source and a device capable of recording the level of the received signal by the antenna under test in a given frequency range are required.

The experimental results indicate the most appropriate configuration for the measuring apparatus among the tested options. When comparing the tested configurations for the apparatus layout in Fig. 7.6, the distribution in Fig. 7.6c produced the most accurate results. Antennas should be placed side by side at one height and separated by a conductive surface to avoid the interaction of antennas. The antenna system should be located at a distance of no more than 1.2 m and no less than 90 cm from the panel. Within this range, the most reliable results were obtained with the least distortion compared to the reference value.

The distance between the chamber walls is not of critical importance. The measuring system is resistant to external influences and undesired wave reflections from surrounding objects. It is believed that the measurement can still be accommodated in a long, narrow room, where reflections from the room walls would be at their maximum. The time required to take the measurement is identical to the standardized measurement in the attenuation chamber. The measurement's duration is influenced by the measurement step, i.e. step in rotation angle and frequency range.

The evaluation of results follows the same methodology as that of standard measurements. When using vector analyzers, data is stored in the touchstone format. Data can be stored in any format, and relevant software can evaluate it. In all the experiments, a proprietary data format was used, listed in a tabular arrangement in a columnar form. For evaluation purposes, the MATLAB environment was used. All data evaluation programs were designed manually to meet the particular requirements of the conducted experiment. By simply adjusting the parameters, the use of MATLAB and LabVIEW ensures the measurement process's significant reliability.

Chapter 9
Conclusion

Since the changes in former Eastern Europe in 1989 and 1990, new trends in regulating production processes and evaluating machines, production processes, and gauges have now impacted our society. In some cases, it has been the result of customer demand and in other cases, it has been to attain a prestigious level, which these methods have certainly facilitated. It is impossible to deal precisely with the analysis of production processes based on the information provided by these methods. It is essential to acknowledge that enhancing the capacity of manufacturing processes, machines, and measuring devices is an economic category, including expensive new machines, active gauges, computer-supported production processes, software equipment, and staff training. A problem may occur when we lack the means to implement, even when we know what we want and need. However, these investments are returned swiftly and therefore, it is profitable to invest in this field. These savings will be prevalent across all domains: This includes savings in the financial area, staff, and material cost due to fewer rejects, and other operational areas. However, the result will also appear in the domain that cannot be purchased but still has a high resale value. The company's image has been growing rapidly in the eyes of its customers. The customer can greatly appreciate it when preventive measures and methods replace the classic final control, which enables high-quality product production rather than control.

The designated goals that we had established prior to the onset of practical implementation in the particular conditions of the unnamed production company have been achieved. For the first time in the history of this company, the production processes, machines, and measuring devices have been evaluated using methods that a serious customer, such as the automotive industry, can easily comprehend. Spatial has been implemented for all production operations to regulate the manufacturing processes, machines, and gauges capabilities. In this manner, we have acquired and will continue to obtain numerous records for conducting a thorough examination of processes, machines and gauges that were previously unavailable.

The work has expanded the area of knowledge on the specific manufacturing processes, machines, and measuring devices. Further analysis is now possible,

M. Krchňák et al., *Advancements in Antenna Measurement*, Lecture Notes in Electrical Engineering 1108, https://doi.org/10.1007/978-3-031-48835-1_9

leading to better quality and more sophisticated products, including bearings in the engineering industry.

SPRINGER NATURE

GPSR Compliance

The European Union's (EU) General Product Safety Regulation (GPSR) is a set of rules that requires consumer products to be safe and our obligations to ensure this.

If you have any concerns about our products, you can contact us on ProductSafety@springernature.com

In case Publisher is established outside the EU, the EU authorized representative is:

Springer Nature Customer Service Center GmbH
Europaplatz 3
69115 Heidelberg, Germany

The manufacturer's authorised representative in the EU is Springer Nature Customer Service Centre GmbH, Europaplatz 3, 69115 Heidelberg, Germany. If you have any concerns regarding our products, please contact ProductSafety@springernature.com

Printed and bound by CPI Group (UK) Ltd, Croydon, CR0 4YY
01/10/2026
02225877-0010